Husserlian Legacies: Themes for the 21st Century

The purpose of this series, *Husserlian Legacies: Themes for the 21st Century*, is to make available to an English-speaking audience the conceptual and methodological resources of Edmund Husserl's works. *Husserlian Legacies* aims to make this material accessible to students as well as researchers in volumes that focus on certain themes and problems that are more focused and shorter in length than the standard Husserliana Collected Works editions. Rather than republishing a lecture or an entire volume, and instead of arranging the material strictly in a chronological order, the *Husserlian Legacies* Series organizes the material systematically with fresh thematic foci. Each volume connects an array of published material in an English translation, uniting them according to themes and problems that are resonant with current philosophical engagements and debates.

Edmund Husserl

Values of Love and Ethical Reflection

Edited by
Sara Heinämaa • Anthony J. Steinbock

Translated by Andrew D. Barrette

Springer

Author
Edmund Husserl
Freiburg im Breisgau, Germany

Editors
Sara Heinämaa
Department of Social Sciences and Philosophy
University of Jyväskylä
Jyväskylä, Finland

Anthony J. Steinbock
Department of Philosophy
Stony Brook University
Stony Brook, NY, USA

Translated by
Andrew D. Barrette
Department of Philosophy
Boston College
Chestnut Hill, MA, USA

ISSN 2731-314X ISSN 2731-3158 (electronic)
Husserlian Legacies: Themes for the 21st Century
ISBN 978-3-031-68697-9 ISBN 978-3-031-68698-6 (eBook)
https://doi.org/10.1007/978-3-031-68698-6

This Springer imprint is published by the registered company Springer Nature Switzerland AG
The registered company address is: Gewerbestrasse 11, 6330 Cham, Switzerland

Series Editors' Introduction

Anthony J. Steinbock in Collaboration with Sara Heinämaa

Beginning with his phenomenological-psychological and logical analyses on the constitution of the concept of number at the end of the 19th century, Edmund Husserl published around a dozen works in German during his lifetime. Regarded by many as a capstone to this publication stretch was his celebrated study of the "crisis" of the European sciences and culture (1936–1937).[1] For such an influential thinker, however, that's only an average of a little more than two works a decade to cover a span of 50 years of prolific thought. This number of lifetime publications gives a highly misleading understanding of Husserl's philosophical contribution. Namely, Husserl belongs to the type of philosopher who strives less to publish, and more to develop their thought by composing new manuscripts. It is thanks to the pioneering and continued efforts of the Husserl Archives in Leuven that authoritative editions of Edmund Husserl's works have been made available in German since the early 1950s, and in English since the 1960s. To date, the catalogue of his writings boasts 43 main titles in the *Husserliana* edition, and 10 others in the *Materialien* collection.

The early image we garnered of Husserl's thought and contributions to philosophy was largely shaped by the availability and reception of the works published in those pre-Husserliana years. We can point to *Philosophy of Arithmetic* (1891), which provided a fertile nodal point of the Husserl-Frege debate on the psychological foundations of mathematics and the ideality of "meaning," and to Husserl's groundbreaking phenomenological analyses presented in his *Logical Investigations* just a few years later in 1900–01, which became a landmark for later phenomenologists, both existentialists and realists.[2] Methodologically, the most

---

[1] The text was titled *The Crisis of European Sciences and Transcendental Phenomenology* (*Die Krisis der europäischen Wissenschaften und die tranzendentale Phänomenologie*) and it was published as volume 6 of the *Husserliana* collection in 1954.

[2] Edmund Husserl, *Philosophie der Arithmetik. Psychologische und logische Untersuchungen. Erster Band* (Halle-Saale: C.A.M. Pfeffer (Robert Stricker), 1981). Edmund Husserl, *Logische*

central of Husserl's lifetime publications was his *Ideas* from 1913, which was planned to consist of three volumes but only materialized in one publication during his lifetime. It is this tripartite whole that allows us to follow Husserl's inquiries through his so-called transcendental methodological turn: The first volume, *Ideas 1* provides a presentation of the new method, whereas the two posthumous volumes, *Ideas 2-3,* utilize this method to perform concrete analyses of sense-constitution and eidetic analyses of the phenomena of nature, life, and humanity. On their basis, he develops a philosophical theory of the relations between the natural and the human sciences.[3]

In addition to these seminal works, Husserl published a series of significant essays and lectures, such as the so-called "Logos" article, "Philosophy as Rigorous Science" (1910–1911), the Vienna Lecture, "Philosophy and the Crisis of European [Humanity]" (1935), and the three Kaizo Articles (1923, 1924), which presented the main results of his phenomenological ethics, axiology, and praxiology.

Compared to what we have at our disposal today in different languages, the initial appreciation of the breadth and depth of Husserl's thought seems both truncated and superficial. While the translation of Husserl's works appeared at a different pace in various languages, the availability of his philosophical analyses and reflections in the English-speaking world was relatively sparse. This is not to say that all the written work by Husserl is equally significant for the shape of this thought. But even in order to understand what might be considered the major threads of his philosophical enterprise, it is imperative to have a sense of the whole.

Moreover, topics that may at first glance seem marginal or secondary can in broader perspectives offer new openings for interpretative work as well as fruitful and systematic starting points for entire new philosophies. We see this for example in case of Maurice Merleau-Ponty, who read selections of Husserl's unpublished manuscripts at the Husserl Archives in Leuven in April 1939, and thereby came to formulate a novel philosophy of perception and embodiment.[4] Another example is offered by Jacques Derrida. A self-proclaimed phenomenologist, Derrida both reinterpreted a few sections that Husserl planned to add to his *Crisis* work on the histori-

---

Untersuchungen. Erster Teil: Prolegomena zu einer reinen Logik (Halle: Max Niemeyer, 1900). *Zweiter Teil: Untersuchungen zur Phänomenologie und Theorie der Erkenntnis* (Halle: Max Niemeyer, 1901).

[3] Edmund Husserl, *Ideen zu einer reinen Phänomenologie und phänomenologischen Philosophie. Erstes Buch: Allgemeine Einführung in die reine Phänomenologie*, ed., W. Biemel (The Hague: Martinus Nijhoff, 1950). Husserliana Vol. 3. *Ideen zu einer reinen Phänomenologie und phänomenologischen Philosophie. Zweites Buch: Phänomenologische Untersuchungen zur Konstitution*, ed., W. Biemel (The Hague: Martinus Nijhoff, 1952). Husserliana Vol. 4. *Ideen zu einer reinen Phänomenologie und phänomenologischen Philosophie. Drittes Buch: Die Phänomenologie und die Fundament der Wissenschaften*, ed., W. Biemel. The Hague: Martinus Nijhoff, 1952. Husserliana Vol. 5. The latter two volumes were originally edited by Edith Stein. Today they are prepared for a new edition by Dirk Fonfara at the Husserl Archives (Husserliana 4/5).

[4] For example, Maurice Merleau-Ponty, *Phénoménologie de la perception* (Gallimard, 1945). Maurice Merleau-Ponty, *The Phenomenology of Perception*, trans., Donald A. Landes (New York: Routledge, 2012).

cal constitution of geometry, and through his careful and creative reading of Husserl's First *Logical Investigation* from another seminar with Paul Ricoeur, instigated a philosophical movement that came to be known as "deconstruction."[5]

The first English translation of Husserl's work appeared as *Ideas: General Introduction to Pure Phenomenology*, by W. R. Boyce Gibson, published in London by George Allen and Unwin Ltd in 1931. This was followed by translations that appeared in the early 1960s and 1970s; these included the *Cartesian Meditations* (1960), *The Idea of Phenomenology* (1964), *Phenomenology and the Crisis of Philosophy* (1965), *Formal and Transcendental Logic* (1969), the *Crisis of European Sciences and Transcendental Phenomenology* (1970), and *Experience and Judgment* (1973). [6] The first work appearing in English translation in the *Collected Works* Series produced by the Husserl Archives was the third book of the *Ideas* (1980), followed by a retranslation of *Ideas I* (1982), and the appearance of an initial rendition of *Ideas 2* (1989).[7]

The early *Anglo-American* Husserl-reception was largely colored by discussions and debates between two alternative "schools" of reading Husserl, the so-called "West Coast" and "East Coast" approaches, with different philosophical starting points, interests, and realms of influence. Whereas the East Coast approach was influenced by deep historical-philosophical considerations (Brentano, Kant, and Neo-Scholasticism), the West Coast readings were more attached to the then current discussions in analytical philosophy, rooted in Frege, but advanced by significant American philosophers, such as Quine, Sellars, Davidson, Kripke, Putnam, and Rorty. Without taking a stand on the question of how well such scholarly divisions may fit into a putative "analytical" and "continental" divide, it is beyond dispute that the picture of Husserl in the *Anglo-American* philosophical theater up to this point was dominated by static and mainly epistemological interpretations of what

---

[5] Jacques Derrida, "Translation and Introduction to Edmund Husserl," *L'origine de la géométrie* (Paris: PUF 1962). See also, Jacques Derrida, *La voix et le phénoméne: introduction au problème du signe dans la phénoménologie de Husserl* (Paris: PUF, 1967).

[6] Edmund Husserl, *Cartesian Meditations*, trans., Dorion Cairns (The Hague: Martinus Nijhoff, 1960); Edmund Husserl, *The Idea of Phenomenology*, trans., William P. Alston and George Nakhnikian (The Hague: Martinus Nijhoff, 1964); Edmund Husserl, *Phenomenology and the Crisis of Philosophy*, trans., Quentin Lauer, (New York: Harper and Row, 1965; Edmund Husserl, *Formal and Transcendental Logic*, trans., Dorion Carins (The Hague Martinus Nijhoff, 1969). Edmund Husserl, *The Crisis of European Sciences and Transcendental Phenomenology*, trans., David Carr (Evanston, IL: Northwestern University Press, 1970). Edmund Husserl, *Experience and Judgment*, revised and ed., Ludwig Landgrebe, trans., James S. Churchill and Karl Ameriks (Evanston, IL: Northwestern University Press, 1973).

[7] Edmund Husserl, *Ideas Pertaining to a Pure Phenomenology and to a Phenomenological Philosophy. Book 3. Phenomenology and the Foundation of the Sciences.* Collected Works, Vol. 1, trans., Ted E. Klein and William E. Poll (The Hague: Martinus Nijhoff, 1980); Edmund Husserl, *Ideas Pertaining to a Pure Phenomenology and to a Phenomenological Philosophy. Book 1. General Introduction to a Pure Phenomenology.* Collected Works, Vol. 2, trans., F. Kersten (The Hague: Martinus Nijhoff, 1982). Edmund Husserl, *Ideas Pertaining to a Pure Phenomenology and to a Phenomenological Philosophy. Book 2. Studies in the Phenomenology of Constitution*, trans., Richard Rojcewicz and André Schuwer, Vol. 3 (Dordrecht: Kluwer, 1989).

phenomenology amounts to, colored by debates on the proper account of the *noema* (the meant as such) and its theoretical status, influenced by neo-Kantian conceptualizations of the transcendental, and shaped by a general equation of transcendental phenomenology with an idealistic theory of human subjectivity. Discussions informed by a one-sided focus on the higher levels of the constitution of objectivity, which were also oblivious to the constitutive role of intersubjectivity, tended to portray Husserl's philosophy as solipsistic and neglectful of the problems of sociality and embodiment, inured to the constitutive role of history, and completely unable to address philosophical problems central to philosophical ethics, political theory, social philosophy, or theory of normativity.

Even if this assessment is painted in broad strokes, it serves to highlight the limitations that the English-speaking world would suffer. Still, there was always a much richer view of Husserl's thought and legacy percolating in philosophical circles that consulted Husserl's works directly. The East Coast vs. West Coast debates also triggered several insightful readings that enriched the polyphony of the reception. These first- and second-generation phenomenologists gave us interpretations of Husserl having a comprehensive yet variated philosophical vision, and helped to parry the notion that he was either a continental version of Frege or a compromised forebearer of Heidegger. As more of his writings have become available, the complexity and scale of Husserl's phenomenological enterprise have come into view, necessitating an expanded presentation of his works in English. What we have is not just an epistemology of mathematics and logic, but a whole philosophical "system" or horizon of thought established on a rich methodology, branching into several subfields: phenomenology of science, social phenomenology, phenomenological ethics, phenomenological semantics and semiotics, transcendental philosophy of logic, and even a renewal of metaphysics.

The purpose of this Series, *Husserlian Legacies: Themes for the 21st Century*, is to make available to an English-speaking audience the conceptual and methodological resources of Edmund Husserl's works. Texts in this Series draw from the rich variety of Husserl's original works and are selected and arranged with the support of the Husserl Archives, Leuven. In an effort to make the material accessible not only to established scholars, but also to graduate and advanced undergraduate students, the volumes appearing in *Husserlian Legacies* are shorter in length than the standard Husserliana *Collected Work* editions and will always be available in a paperback. Further, rather than republishing a lecture or entire volume from the *Husserliana* edition, the *Husserlian Legacies* Series organizes the material systematically with fresh thematic foci. Finally, instead of arranging the material strictly in a chronological order, each volume of the *Husserlian Legacies* Series connects an array of published material in an English translation, uniting them according to themes and problems that are resonant with current philosophical engagements and debates.

This first volume of *Husserlian Legacies: Themes for the 21st Century: Values of Love and Ethical Reflection*, focuses on hitherto unknown dimensions of Husserl's philosophical thinking. The volume helps us both to appreciate Husserl's phenomenological analyses of value and love and to gauge their systematic role in his reflections on several related topics: his concept of philosophy, the sciences and the arts as vocational tasks, the role of emotions in ethics, his idea of the community of love, and his personalistic reformulation of the categorical imperative. The texts gathered in this volume thus have far-reaching implications, ranging from axiology and individual and social ethics to the theory of emotions and the philosophy of science.

Future volumes in development and planning stages include: Volume 2: *Vocations and Teleology*; Volume 3: *Normality and Abnormality in a Transcendental Aesthetic*; Volume 4: *Norms and Ethical Normativity*; Volume 5: *Philosophical and Phenomenological Anthropology*.

Husserl's Investigations into Love and Values of Love: Axiology, Praxeology, and Ethics

Sara Heinämaa in collaboration with Anthony J. Steinbock

This first volume of *Husserlian Legacies: Themes for the 21st Century* focuses on the theme of love, which is usually not associated with Husserl's philosophical work and is routinely assumed to be marginal to or outside of his phenomenological investigations. However, in Husserl's mature and late research manuscripts from the 1920s and 1930s, we find profound new analyses of the emotion of love and original discussions of its importance to axiology and ethics, both individual and social.[1]

The volume throws light on these hitherto unknown dimensions of Husserl's philosophical thinking-in-progress and allows us to appreciate their role in the establishment of some of his well-known philosophical results, most importantly, the analogy between ethics and logic, the personalistic reformulation of the categorical imperative, the idea of the community of love, and the understanding of philosophy and the sciences as vocational tasks.

The texts gathered in this volume thus have far-reaching implications for several subfields of contemporary phenomenology and philosophy more generally, from axiology and individual as well as social ethics to philosophy of emotions, epistemology, and philosophy of science. The collection thus resonates with Husserl's statement from 1935: "Love in the genuine sense is one of the main problems of phenomenology, and not in abstract detail and isolation, but as a universal problem."[2]

---

[1] I am grateful to John Drummond and Julia Jansen for their insightful comments on and very helpful suggestions for an earlier version of this paper.

[2] Hua42, 524; cf. Sowa and Vongher 2013, ciii, note 4 (starts cii).

1. Ethics as a Theoretical Science

Husserl's late discourse on love and values of love (*Liebeswert*) is part of his phenomenological account of intentional emotions and their passive grounds in non-intentional feelings, drives, and sensations.[3] This account is not a branch of Husserl's phenomenological psychology or a division of his phenomenological anthropology,[4] but is a central development and extension of his philosophical theory of persons and personal wholes, instituted in the 1920s in the second volume of *Ideas*, systematized in *Cartesian Meditations* and related manuscripts, and further elaborated till the very last writings in the 1930s.[5]

Husserl's philosophy of persons is transcendental-eidetic in its methods. It illuminates the actual lives and coexistence of human beings,[6] but proceeds to investigate into the conditions of possibility of all conceivable persons *in toto*. The most important exemplars of persons in this reconceptualization are not among infantile, animal, artificial, extraterrestrial, or divine subjects,[7] but are the open-ended organizations of individuals that Husserl calls "personalities of higher order" (*Personalitäten der höheren Ordnung*). Such higher order personalities include nations, cultures, states, federations, churches, and religious communities, but also the universal communities of the sciences and the arts.[8] Even social classes, genders, races, and other

---

[3]For Husserl's first sketches of a theory of emotions in his *Logical Investigations* (*Logische Untersuchungen* 1900–1901; Hua19/2), see Jardine 2020; cf. Mayer and Erhard 2019.

[4]E.g., Luft 2010.

[5]The second volume of Husserl's *Ideas* is available in two different editions, one composed by Edith Stein and finalized by Ludwig Landgrebe (Hua4; Hua5), and the other, very recent one edited by Dirk Fonfara (Hua4/5). *Cartesian Meditations* was long known only in French translation, *Meditationes Cartesiennes* (1931), and came out in German only in 1950 (Hua1; cf. Heinämaa 2021). Husserl continued to work on his theory of intersubjectivity and monadology until his very last years. The results of these investigations were published in 1973 in the so-called intersubjectivity volumes XIII–XV of the *Husserliana* collection (Hua13–Hua15). For Husserl's philosophical theory of persons, see Heinämaa 2007; Moran 2009; Jacobs 2010; Hahn 2012; Caminada 2016; Moran 2017; Jacobs 2020; Heinämaa 2022a; Jardine 2022; 2024/2025; for the personalistic foci and grounds of his ethics, see Melle 2002; Donohoe 2004; Melle 2007; 2008; Crespo 2015; Heinämaa 2014; Ferrello 2016; Vadevelde 2016; Obsieger 2017; Breuer 2019; Cavallaro and Heffernan 2019; cf. Hart 1992; 2021.

[6]The German word "der Mensch" (*the human being*) that Husserl uses when developing the anthropological implications of his phenomenology of personhood is syntactically a masculine noun. Its grammatical gender in German requires that it must be indicated by masculine pronouns "er" and "sein" (genitive), and not by feminine or neuter ones. So, Husserl's discussion of "der Mensch" and "seine" capacities is devoid of androcentric, patriarchal, and sexist connotations, as is also his discussion of "die Katze" (*the cat*) and "ihre" movements (Hua4, 175–176/185; Hua4/5, 180).

[7]We find discussions of non-human persons, for example, in the second volume of *Ideas* (Hua4; Hua4/5) and the so-called intersubjectivity volumes (Hua15).

[8]E.g., Hua1, 137–138/107, 149–150/120–121, 157–161/129–133; Hua4, 181–182/191, 194–197/202–206, 243–245/255–256; 315–316/328–329, 360–361/370–371; Hua4/5, 195–196, 243–244, 298, 525, 544, 561–563, 586; Hua15, 421, 471–472, 478, cf. 461–479, 510–513; Hua27, 22; Hua42, 223, 284; cf. Carr 1987; Moran 2013; 2014; Carr 2017; Moran 2017; Caminada 2019; Miettinen 2020.

so-called "identity-categories" may be analyzed as personalities of higher order insofar as they are constituted by their members in communicatively shared acts of valuing and willing and are not simply produced by socio-historical forces or circumstances.[9]

The ethical ideal of such higher-level personalities for Husserl is the community of love (*Liebesgemeinschaft*) in which all members are actively united by symmetrical and transitive relations of loving.[10] By this regulative idea, Husserl reinterprets the Christian notion of neighborly love and concretizes his discussion of a non-revelational rational faith (*freies und rationales Glauben*).[11] The mereological ground of this "renewal" of Christianity lies in Husserl's Leibnizian social ontology, that is, his monadology,[12] but the idea is also informed by his interests in Stoic and Kantian cosmopolitanism.[13]

For systematic methodological purposes, it is important to realize that Husserl's discourse on personal love and love between persons primarily contributes to his transcendental-eidetic inquiries into the essential structures of all possible value-intending and willing.[14] When proceeding in his inquiries into love, he observed that several inherited presumptions and habitual convictions about emotions and the values intended in them are untenable in light of the demands of his transcendentally radicalized critique. He thus came to think that the cornerstone of his rational scientific ethics—formal axiology—must be reworked. More precisely, the analytic results of Husserl's investigations into personal values of love urged him to

[9] Cf. Flynn 2012; Drummond 2020a; 2023.

[10] E.g., Hua14, 172–175; Hua42, 301–302, 314–318, 390; 408–409; cf. Buckley 2000; Melle 2002; Crespo 2018; 2022. The concept of community of love also figures centrally in Max Scheler's phenomenology that Husserl knew through Scheler's early publications, most importantly, *Formalism in Ethics and Non-formal Ethics of Values* (*Der Formalismus in der Ethik und die materiale Wertethik* 1913/1916). Scheler used the concept for systematic purposes in his investigations of differently structured human communities and types of intersubjectivity. In his usage, communities of love are characterized by co-responsibility for all members—both individual and collective persons—constituted in and through mutual loving. Such communities and only such communities can, in Scheler's analysis, usher in a genuine solidarity through loving. On this basis, he distinguishes communities of love from three other forms of human sociality: (i) *life-communities* (*Lebensgemeinschaften*) which are focused on the value of life and which entail responsibility for the communal life as a whole, (ii) *societies* (*Gesellschaften*) in which the value of individuals is recognized over the community, requiring implicit or explicit contracts, and (iii) *the mass or herd* (Scheler [1913/1916] 1973; [1931] 1976/1977; cf. Steinbock 2014; Zahavi 2018).

[11] Hua27, 66–67; cf. Hua29, 109; Hua42, 242, 254–255, 390.

[12] Hua1, 166/139, 139/108, 166–171/139–144, 182/150–151.

[13] E.g., Hua42, 236, 254, 414; Hua7, 17ff.; cf. Leitlande 2022; Heinämaa 2020; Miettinen 2020; Cavallaro and Heffernan (eds.) 2022; Ferrarello 2023.

[14] The studied eidetic structures include both formal and material ones.

reconsider the adequacy of the axiological principles of his early ethics.[15] These values are disclosed by subjects, but their subjective disclosure does not make them relative; rather, they disclose themselves as absolute and place unconditional obligations on persons. They are not objectifiable[16] or comparable with other values and thus they defy the axiological laws which organize values in hierarchies of priority and preference.[17] Finally, the demands that these values place on our actions and

---

[15] In his early lectures on ethics, *Vorlesungen über Ethik und Wertlehre 1908–1914* (Hua28), Husserl put forward a set of formal axiological laws which all depended on the idealizing assumption that all values are objectifiable equally and thus mutually comparable. When his investigations into persons and personal values progressed, he came to question the tenability of his assumptions about universal objectification. He proceeded in his reflections by conceptualizing love-values which are not thus objectifiable, and at the same time argued that persons themselves—as unique sources of meaning and value—are not objects and not objectifiable in the manner in which things and other realities are. This limitation is due to the open-ended and dynamic character of the essences of persons (e.g., Hua5, 17–18; Hua14, 14). While explicating his social philosophy in the second volume of *Ideas*, he already contended: "The persons who belong to the social association are given to each other as 'companions,' not as opposed objects but as counter-subjects who live 'with' one another, who converse and are related to one another, actually or potentially, in acts of love and counter-love, of hate and counter-hate, of confidence and reciprocated confidence, etc." (Hua4, 194/204). In the research manuscripts (ca. 1925), he writes: "It belongs to the 'essence' of each personal individual, each ego [*Ich*], that the individual has a field of personal decisions, personal love, personal ought [*Sollen*]. Of course, talking about essence here is specific, since the ego does not have an essence like matters [*Sache*] have" (Hua42, 355; cf. Hua4, 270–271/283). However, Husserl did not completely abandon all of his formal axiology (cf. Rinofner-Kreidl 2022). He came to see its serious restrictions and realized the theoretical importance of delimiting such an axiology so that it does not absolutize itself as "the ethics" or "the morality." (These last points are debatable, however: several scholars argue that Husserl's late reflections on love led him to question his early axiology as a whole, e.g., Melle, Drummond. In contrast, I believe that he still saw his early axiology as descriptively valid for capturing certain aspects of value-experiences, but merely in situations in which all relevant values are commensurable.)

[16] On objectifying and non-objectifying acts, see Bernet 1994; Drummond 1995; Benois 2002; Melle 2019; Mayer and Erhard 2019. On Husserl's absolute, see Hart 2006; cf. Marosan 2020.

[17] A comparison to Scheler's ethics is illuminative, since there are certain parallels but also important differences between these two phenomenological approaches (Rinofner-Kreidl 2013; Steinbock 2014; Obsieger 2017; De Monticelli 2021; Rinofner-Kreild 2022). According to Scheler, values are ordered in terms of their valences and ranges. The order is "objective" for the experiencing person, but this does not mean that values are objects (*Gegenstände*) or given to emotionally inert observers in a disinterested, third person perspective. Rather, in Scheler's account, values are given through specific acts with a cognitive dimension all their own, and as such they have a depth or height by virtue of the kind of valence that they carry. It is this valency that orders values for feeling persons in relation to other values. So, values and orders of values necessaril appear to persons, and they also have to be given through some media (in general, "feelings" as value-cognitive forms of intentionality) that conveys them to valuing subjects. What we love, for example, are particular persons as bearers of absolute "spiritual" values, not value or spirit in abstraction. Still, values exhibit a kind of independence; they are not reducible to the subjects who experience them nor to their bearers. They only appear in what Scheler calls "ranges," not by individual whim, but by features that give themselves in experience. On this basis they are experienced as deeper or more superficial, as higher or lower, in relationships of founding and shareability. It is these relations of founding and shareability that participate centrally in the constitution of value hierarchies for Scheler. Thus, the ordering of values concerns their intrinsic structures of appearing and being

forms of agency can be disobeyed, but no such disobedience can lead to or restore any inner balance in us. Instead, each such failure leaves us in permanent sadness and grief.[18] So, Husserl's late studies of the emotion of love and his conceptualization of values of love (as correlates of acts of genuine loving) challenged several assumptions on which his own early axiology rested. This forced him to rework his ethics and abandon some of its basic assumptions about valuations and values.

Already in the early lectures on ethics from 1908 to 1914, Husserl clarified the character of ethics as a science by comparing it with logic. Both sciences deal with ideal objectivities of some sort—one with the essential structures of value-meanings and the other one with propositions and judgment-forms—and both are thus *a priori* eidetic disciplines, open to transcendental-critical reflections (*Besinnung*) on their constitutive foundations. The ideality of ethics and logic implies that all empiricist—psychologistic, naturalistic, anthropologistic, and historistic—accounts of them must be rejected. No empirical science, bound by realities of some sort (the psyche, nature, humanity, culture or history), can alone make sense of ideal existence and relations or the ideal generalities and laws that govern them.[19]

The other crucial implication of the analogy is that neither logic nor ethics is a normative practical discipline (*Kunstlehre*), but both are fundamentally theoretical undertakings.[20] As such they do not simply answer questions about *doing* or what *ought to be done* but necessarily ground their normative statements on descriptive and explicatory statements about *being*, more precisely, ideal being of some sort

founded. Finally, through loving, the bearer of value can be given to the loving person outside of all value hierarchies. Thus, Scheler distinguishes acts of loving from acts of preferring, and argues that only the latter are comparative, whereas acts of loving operate in absolute terms. Ultimately, he contends, choosing is founded in preferring, which in turn, is founded in loving.

[18] Heinämaa 2020.

[19] For Husserl's analogy between ethics and logic, see Drummond 1995; Rinofner-Kreidl 2022.

[20] The German term "Kunstlehre" that Husserl uses in his typification of the sciences is translated into English often by the phrase "technological science" (e.g., Findlay's translation of *Logical Investigations*; cf. Bernet 2002; Moran 2005). This solution is etymologically illuminative, since it emphasizes the ancient Greek roots of Husserl's conceptualization of the practical sciences as sciences of arts and skills (*λόγοι* from *λόγος*; *τέχνες* from *τέχνη*), in distinction from the theoretical sciences of being (actual, possible, real, ideal). However, semantically considered, the English term "technological" is misleading in the context of Husserl's epistemology, since the term has distanced far from its Greek origins and ultimately separated from them. Today (as in Husserl's time), "technological" primarily refers to those human practices which utilize *mechanical instruments* or *computational machinery* and which thereby differ drastically from the age-old arts and skills (Hua42, 231–233; cf. Buchanan). Ethics, aesthetics, and pedagogy are not among technological practices in this late modern sense—even if they may be susceptible to formalizations and thereby also to computational mechanization (Hua6). So, it is not "technological sciences" in the contemporary sense of the English word that are at issue in Husserl's discussion of *Kunstlehren*. Husserl does not define *Kunstlehren* conatively by their goals and means but defines them logically by the types of laws (and judgments or statements) that they issue; whereas the laws of the theoretical sciences are descriptive (or explicatory) in kind, the laws of *Kunstlehren* are prescriptive (Hua18, 59/37–39; Hua37, 13–24; cf. von Wright 1963, 7–10). For these logical reasons, I have decided to render the German "Kunstlehre" into English as "practical science" and "practical discipline." Another alternative would be "applied science."

(senses, meanings, values, propositions, etc.). This is Husserl's main critique against his teacher, Franz Brentano, who treated logic and ethics equally as normative practical sciences.[21] Analogous arguments can be put forward with respect to other normative sciences as well, most importantly aesthetics but also pedagogy, linguistics, and political theory. Already in *Logical Investigations* from 1900 to 1901, Husserl formulates this general insight into an epistemological thesis, which he then backs up by arguments in the case of ethics in his writings on axiology and praxeology:[22] "Every normative discipline demands that we know certain non-normative truths: these it takes from certain theoretical sciences, or gets by applying propositions so taken to the constellation of cases determined by normative interests."[23]

More concretely, ethics deals theoretically with (i) *the value/disvalue* of volition and *the correctness/incorrectness* of action and (ii) *the emotions* that motivate good/bad willing and correct/incorrect action.[24] Phenomenologically considered, its main interest lies in the eidetic structures of these phenomena, and therefore it must ultimately be a theoretical science with ideal objects and *a priori* methods. However, from theoretical ethics a normative ethics can be "derived" by certain transformations.[25] Such an ethics tells us how to behave and what to do on the basis of an *a priori* and independent eidetic account of what goodness, beauty, correctness, and other value-phenomena mean.

The theoretical core of ethics entails two mutually related parts, an axiological part that deals with the *goodness/badness* of the willing and the *correctness/incorrectness* of actions and motivations, and a praxeological part that deals with the *volitions* that can be good or bad and the *actions and motivations* that can be correct or incorrect. [26] The former draws from the general theory of valuations and values, and the latter from the general theory of volitions and their goal-objects. However,

[21] Brentano [1889] 1969; [1952] 1954.

[22] Hua28, 37ff.; Hua37, 20–23, 31–32, 271; cf. Hua17, 121/136–137; Hua27, 87–88, 118–120; Hua43/2–3.

[23] Hua18, 61/39; cf. Hua17, 27–28/31–32; Hua24, 27–28/27; Hartimo 2018; 2020.

[24] In more detail, one can say that ethics deals theoretically with (a) the positive/negative value of the ends at which volitions aim and the rightness or wrongness of actions as well as (b) the emotions that motivate willing and action. I owe thanks to John Drummond for this specification which emphasizes that what we will are eventually the goals of actions and that we are able to will both good and bad goals.

[25] Hua18, 59/38; Carta 2021; Heinämaa 2022b.

[26] In Husserl's analysis, volitions or acts of will—so, decisions, resolutions, and actions—are not emotions, desires, or beliefs of any sorts nor intentional complexes combining emotions, desires, and/or beliefs. Rather volitions are *sui generis* conscious acts with their own intentional structures, own manners of directedness, own intentional objects (goals, means, purposes of various sorts), and own temporal forms and dimensions (Nenon 1991; 2009; Mulligan 2004; Heinämaa 2022b; Spano 2022a; 2022b; cf. Salice 2018). They are similar to desires in having a necessary futural orientation, but they differ from desires in being differently bound to and interested in reality and actual being (Heinämaa 2023). Like emotions, they are founded on doxic acts of some sort but not in a similar way and not in parallel with emotions.

the praxeological part of theoretical ethics is dependent on its axiological part,[27] since an account of value and correctness is necessarily needed in order to delineate and identify the main subject matters of ethics: good will and correct action.

The core idea of this analogy between ethics and logic is retained by Husserl up until his last reflections,[28] but its main framing is changed. Namely, when Husserl proceeded in his investigations into love, he realized the strongly motivating force of this emotion in all agentive life and came to see the very specific intentional and egoic features of some of its variants.[29] This shook his confidence in the adequacy of the formal axiological laws put forward by his early ethics: some of the results of his new analyses suggested that the values that are disclosed by love may not be objectifiable in the same manner and to the same extent as those disclosed by other emotions. The result seemed to hold for several prominent types of love, despite the fact that that they are called by different names in our philosophical tradition, such as *eros*, *storge*, *philia*, *philautia*, *agape*, *caritas*…

2. Emotions as Complexes of Intentional Acts

In Husserl's account, all moral teachings and all normative ethics depend in their senses on experiences of correctness, value, and valuation.[30] Emotions are therefore crucial for such disciplinary undertakings, since they are our original intentional manners of relating to values. More concretely, they are the specialized forms of experiencing by which values are given to us and by which we can organize our actions and our lives into axiologically and practically meaningful wholes. So, it is emotions that disclose to us the values of sensory pleasures, economic success, benevolence, self-realization, clearness and distinctness, justice and equality, as well as the disvalues of violence, cruelty, and stupidity. Moreover, emotions allow us to participate by our own individual actions in the promotion and realization of ideal values, such as beauty and goodness, or the value of truth. Emotions are thus necessary at least in three axiological respects: for intentional relatedness to values, for the organization of actions and lives according to values, and for the realization of ideal values in the material world.

The German term "Gemüt" that Husserl uses for emotions in general is often, as also in this volume, translated into English by the term "heart."[31] The translation has

---

[27] Cf. Crespo 2015, 702.

[28] The analogy is motivated by Husserl's insight into the unity of reason, despite its three different manifestations in thinking, valuing, and willing (cf. Rinofner-Kreidl 2022). On Husserl's early discussion of the analogy between ethics and logic, see Drummond 1995.

[29] Melle 2002; 2007; Loidolt 2012; Crespo 2018; Drummond 2018; Heinämaa 2020.

[30] Cf. Drummond 1995; Obsieger 2017.

[31] I follow here the alternative practice, established by Cairns, Dalhstrom, and others, of rendering "Gemüt" by "emotion," since it makes systematically more sense to me than the metaphorical "heart" in respect to Husserl's overall philosophical aims (cf. Heinämaa 2024).

many benefits: It helps highlight Husserl's critique of Kant's moral philosophy and thus resonates well with the insights of his contemporary Kant-critics, most importantly his teacher Franz Brentano and his fellow phenomenologists Edith Stein, Max Scheler, and Nicolai Hartmann.[32] Furthermore, since the term "heart" figures prominently also in the philosophies of Husserl's later Christian followers, such as Dietrich von Hildebrandt, Jean-Luc Marion, Robert Sokolowski, and James Hart, it allows us to track several important developments in 20th-century phenomenological ethics.[33]

All these philosophers contend that emotions, especially certain forms of love, have a crucial role in the ethical lives of persons as motivating forces. Thereby they all join in challenging the dominant deontological paradigm according to which moral actions can properly be motivated merely by respect for the moral law by pure practical reason.[34] The translation of "Gemüt" by "heart" captures well this common insight that united many 20th-century phenomenologists despite their other differences, Husserl included: they all rejected the Kantian picture of a head (practical reason) that guides a moral person independently of all affections of the heart (emotion), be they correct or incorrect.[35]

However, the term "heart" also carries certain connotations that prove problematic in the context of Husserl's philosophical system. The metaphorical contrast head/heart suggests a set of habitual philosophical dichotomies, such as reason vs. emotion, intellect vs. feeling, judgment vs. valuation. Husserl believes that such dichotomous conceptual tools block eidetic analyses of intentional experiences instead of promoting them.[36] For him, reason is not one of the (human) capacities, but the process of evidencing, characteristic of all types of intentional acts, emotive as well as intellectual, axiological as well as theoretical.[37]

More precisely, Husserl accepted Brentano's insight into the insufficiency and abstract formality of the Kantian paradigm of morality, but he did not accept Brentano's typology of intentional acts. In his view, Brentano's contrastive distinctions—judgment/emotion and love/hatred—were simplistic and misleading, even if

[32] Brentano [1952] 1954; Scheler [1913] 1973; [1913/1916] 1973; cf. Luft 2010; Salice 2018; Drummond 2021b; Rinofner-Kreidl 2022. For Scheler's use of the metaphor of the heart, see Gubser 2014; Josset 2021; Enquist Källgren and Vendrell Ferran 2022. For the Pascalian roots of this metaphor, see Ricœur 1994.

[33] Von Hildebrand [1965] 1970; Marion [1986] 1999, Ch. V, 293–380/277–352; Sokolowski 2008; Hart 2009; cf. Ricœur 1994; Flynn 2010; Steinbock 2014; Vandevelde 2016; Steinbock 2018; 2021; Gschwandtner 2022.

[34] The only emotion that has an indispensable role in the Kantian paradigm is *respect* which is felt for the moral law as such and for the free agents who autonomously submit themselves to this law by acts of reason (cf. Drummond 2006). Other emotions may have pedagogical or heuristic functions but none of them is necessary for the motivation of moral agents (Sorensen and Williams (eds.) 2018). For a defense of Kant against Husserl's alleged "misreadings," see Rinofner-Kreidl 2010. For Husserl's indebtedness to Fichte (and Kant), see Drummond 2018; Rizo-Patrón 2019.

[35] Kant [1797] 2009, 6:480.

[36] Cf. Drummond 1995; Peucker 2004; Obsieger 2017.

[37] Hua42, 240; cf. Hua3, 333ff./326ff. (Hua3/1, 314ff.); Hua17, 120–121/135–137.

Brentano argued that all these forms of intending contribute to concrete experiences and are grounded in presentations (*Vorstellungen*). To liberate himself from traditional dichotomies and pay proper heed to the plurality of experiencing, Husserl put forward a more complex distinction between three kinds of intentional acts: doxic, axiological, and conative.[38] A few words about these basic analytical tools are needed to make sense of Husserl's mature philosophy of the emotion of love.

Doxic intentionality is the form of intentionality that is responsible for the positing of the many senses of being, e.g., being present or absent; real or ideal; certain, probable, likely…; being human, animal, living, female, male… It is predominant, for example, in cognitive, perceptual, recollective, and image-forming experiences.[39] Its passive grounds and materials are to be found in sensations of various sorts, tactile, kinesthetic, visual, auditory, and in the synesthesia that unites specific forms of sensibility. *Axiological intentionality* posits values, for example, the pleasurable, beautiful, correct, valid, just, good… Axiological acts are operative in value-intuitions or "valueceptions" (*Wertnehmen*),[40] evaluative judgments, emotions, and intentional feelings.[41] Their passive grounds are in non-intentional feelings and instinctive affections. Some axiological posits are epistemologically central, while others are crucial to ethics and morality. Epistemic values entail, for example, the values of clarity and distinctness and the fundamental value of truth;[42] moral values entail the various values of virtuous life, the values of correct action and beautiful behavior, and the governing value of the good will. Furthermore, the aesthetic values of beauty and sublimity figure prominently in artistic practices and arts, but Husserl also argues that beauty has an important ethical function in the valuing of personal styles of becoming.[43] Finally, *conative or practical intentional acts* posit goals and means. They are predominantly operative in volitions, decisions, resolutions, and actions which are grounded in the passive consciousness of drives and

---

[38] E.g., Hua28, 156–157.

[39] Hua23.

[40] By the neologism "wertnehmen," Husserl refers to original axiological acts that relate us to values, prior to and independently of logical categorization and distinction. He uses this term in parallel with the standard German "wahrnehmen" (*to perceive*) that refers to perceptive acts which also are able to operate independently of logical articulation. On the basis of this analogy, the derivative noun "Wertnehmen" is translated into "valuception." However, Husserl's analysis demonstrates that both modes of original intending—*wernehmen* and *wahrnehmen*—are susceptible to logical categorical acts of objectification.

[41] In Husserl's analysis, emotions do not conceptually grasp, represent or apprehend objects, or anything, independently of objectifying doxic acts (e.g., Hua3, 82–83/76–77, 264–265/274–275; Hua3/1, 76–77, 263–265; Hua43/2, 307–309, 330–338, 433–436; cf. Drummond 1995; Jardine 2020). They are best described as axiological acts by which we *take position* vis-à-vis our non-intentional feelings and sensations.

[42] E.g., Hua17, 53–55/60–62.

[43] Hua42, 337; Hua43/2, 113–114, 234, 247–257, 507–535; cf. Hua4, 270–271/283.

instincts. But also certain long-term emotions are informed by conative manners of intending.[44]

In Husserl's analysis, each of these act-types has relations of dependence with other ones but they are not all similarly founded on presentations, as Brentano argued. Most importantly, the founding of judgments differs significantly from that characteristic of emotions.[45] Moreover, among the non-doxic acts, only axiological acts are *directly* founded on doxic acts.[46] Simple presentations can serve as such foundations but also more sophisticated perceptions, cognitions, imaginations, or memories. In contrast, conative acts of willing are directly founded on axiological acts and only via their axiological foundations further founded on doxic acts.[47] To get a clearer picture of Husserl's theory of emotions and the central position of the emotion of love in it, we need to look more closely into the variance of the emotive objects and into the different manners in which the three types of intentional acts can combine in concrete emotive experiences.

The first thing to notice is that Husserl's thesis on the doxic foundedness of axiological acts does not state that all axiological acts would *actually* be established on the top of some doxic acts. Rather than claiming anything about actual relations between acts, the thesis concerns *possibilities*.[48] The main idea is that each value-intending experience allows a conversion into a non-valuing one, but the contrary does not hold.[49] For example, the emotive delight in the gorgeous gallop of a horse can be converted into an evaluative judgment about the movements of the animal;[50] but the judgment that 2 + 2 = 4 cannot be converted into any emotion or any valuation whatsoever. The point in the objectification of emotions and feelings is to enable their discursive explication in speech and thinking.[51]

---

[44] For the intentional and temporal structures of volitions and conative intentionality in general, see Heinämaa 2023.

[45] Hua43/2, 334–335, 365–367; cf. Hua3, 284–285/276–277 (Hua3/1, 266–267). On Husserl's understanding of the relations of founded-founding, see Drummond 2003; 2013; Rinofner-Kreidl 2013.

[46] E.g., Hua28, 72, 252, 255, 267–268, 338; Hua37, 274; Hua43/2, 334–337, 341–344; cf. Hua3, 281–291/274–282 (Hua3/1, 263–272), 328/321 (Hua3/1, 309); Hua4, 8–9/9–10; Drummond 2009; 2013; 2019; Jardine 2020a.

[47] Hua28, 126–127; Hua37, 24, 215, 293, 325–326, cf. 215–216. Cf. Rinofner-Kreidl 2022, 252. Scheler accepts part of Husserl's analyses of the relations of founding between objectivating and non-objectivating acts, but he argues from 1913 onward that emotional acts—and in particular acts of loving—are foun*ding* for presentational acts, perceptual as well as judicative (Scheler [1913/1916] 1973; [1926] 2021).

[48] Hua3, 291/282 (Hua3/1, 272); cf. Hua42, 241.

[49] E.g., Hua3, 289–290/280–282 (Hua3/1, 269–270), 299/290 (Hua3/1, 280–281); Hua4, 8–9/9–10; Hua28, 119; cf. Drummond 1995; Rinofner-Kreidl 2013.

[50] Hua43/2, 15–17; cf. Hua4, 8–9/9–10; 187–188/197–198.

[51] E.g., Hua3, 311/301–302 (Hua3/1, 292); Hua19/1, 332–333/65–66; Hua28, 63; Hua43/2, 335; cf. Drummond 1995; Benois 2002; Jardine 2020. The function of objectifying acts can be characterized in logical (Fregean) terms by saying that they allow the *isolation* of the noematic sense from the noetic act and thus make possible the intending of the isolated sense by other acts (cf. Drummond 1995; Benois 2002).

The second thing to notice is that the doxic foundation of emotion does not need to be a reality—a real thing, event, or process—or entail any real being of any sort.[52] Many important types of doxic acts have non-existent objects as their noematic correlates. Such acts include, for example, memories and phantasies, both of which can found emotions. Neither types of acts need to be directed at any actual real being whatsoever in order to perform their intentional functions. Instead, both relate to non-existing entities, but in different manners: a memory ties the recollective subject to some past being that was given in actual experience earlier but is now absent, lost, or destroyed; an act of phantasy in contrast connects the subject to a merely possible being.[53] Moreover, also purely ideal objects can operate as intentional correlates of doxic acts. Finally, intentional acts can also be founded on mere appearances as such without activating any further positing of the sense of being—real or ideal, actual, or possible.[54] All these non-real objects can serve as foundations of valuations. We can, for example, be struck by the beauty of a rainbow or Northern lights (appearances) and admire the elegancy of a mathematical proof (ideality).

So, emotions can base their valuations on various kinds of formations: persons, actions, things, events, appearances, actualities and possibilities, realities and idealities. Unlike actions they are not bound to realities. Thus, they can be said to be relatively "openminded" in respect to the doxic foundations that they require.

This implies that if we want to study phenomenologically the emotion of love, we must not restrict ourselves to actual or real objects merely. Imaginary objects and ideal ones are equally important for the theorizing of the phenomena of love. Even if we ourselves would not be able to love God or any divinities—or even understand that someone can love such beings—we should still not exclude such experiences from our phenomenological analysis of the emotion of love. So, the mystical-erotic emotions and feeling-states that Teresa of Ávila had vis-à-vis Christ[55] as well as the sentiments that Daniel Paul Schreber had in respect to God,[56] or the ones that John Hinckley Jr experienced towards Jodie Foster, are all relevant to our analysis, and equally relevant as our own ordinary emotions toward our nearest ones.[57]

Furthermore, completely fictional cases as well as distantly historical ones are informative to phenomenological investigations into emotions in general and love in particular. The historical love-affair between Abelard and Héloïse or the imaginary experiences lived through by the Beauty and the Beast or Anna Karenina and Vronsky must not be excluded simply on the grounds that they have no actuality or

---

[52] Hua37, 189–190; Hua43/2, 309, 337–338.

[53] Hua23; cf. Jardine 2024.

[54] Hua43/2, 248–258, 309, 337–338; cf. Hua37, 189–190.

[55] Stein [1935/1936] 2002, 369–373; cf. Steinbock 2007/2009.

[56] A famous case of delusory love or erotomania.

[57] As Ingmar Bergman contends through one of the characters of his film *Through a Glass Darkly* (*Såsom I en spegel* 1961): It is not a matter of a special kind of love, but all kinds, "the highest and the lowest, the most absurd and the most sublime. All kinds of love."

reality.[58] Finally, the whole qualitative spectrum of loving has to be taken into account, not just mystical-erotic or romantic cases. This entails parental love, patriotic love, one-sided love, and love for ideal objects, just to mention a few variants central to Husserl's late reflections. Thus, the general guideline that Husserl formulated for his phenomenological science and the eidetic sciences *in toto* is especially suitable for intentional studies of the structures of love:

> Extraordinary profit can be drawn from the offerings of history, and in even more from those of art, and especially from poetry, which are, to be sure, fruits of imagination but which, in the originality of new formations, the abundance of detailed features and the unbrokenness of motivation, tower high above the products of our own phantasy (…) Thus if one is fond of paradoxical phrases, one can actually say, and if one means the ambiguous phrase in the right sense, one can say in strict truth, that *"fiction" makes up the vital element of phenomenology as of every other eidetic science*, that fiction is the source from which the cognition of "eternal truths" is fed.[59]

Husserl's investigations into emotions do not proceed by introspections or first-person reflections merely.[60] Rather they require and utilize his methods of eidetic variation, radical self-reflection (*Selbstbesinnung*), and ultimately also transcendental purification. To be sure, such investigations can communicate fruitfully with both empirical studies of the affective lives of human beings and with analytical-philosophical inquiries into our emotive lives but, all the same, they must leave behind both empirical generalizations and conceptual analyses in order to answer constitutive questions about the open universe of possible forms of loving.

Most concrete experiences entail all three types of intending—doxic, axiological, and conative—but even if these types are thus entangled in our intentional lives, we can conceptually distinguish between them in terms of their characteristic forms of directedness and temporality. Moreover, in most concrete experiences, one intentional-temporal type of intending unites the whole by founding the others. The regard or focus of the axiological act, however, is not in the "subject matter" (thing, event, situation) of the founding act as it presents itself independently of value-intending; the founding subject matter is meant, but meant insofar as it is articulated by the axiological act itself.[61]

Husserl argues that full-grown love, love in the pregnant sense of the word, is governed by axiological acts of value-intending but also entails a *conative* act of

[58] E.g., Hua43/2, 330; cf. Rinofner-Kreidl 2016; Jansen 20180.

[59] Hua3, 163/160 (Hua3/1, 148), translation modified; cf. Hua4, 120/127; 266–267/278–279; Jansen 2020.

[60] Cf. Drummond 1995; Drummond and Rinofner-Kreidl 2018, 1–2.

[61] E.g., Hua3, 290/282 (Hua3/1, 272); cf. Jardine 2020.

committing oneself to caring for the loved one as well as for one's capacity to love.[62] Thus analyzed, full-grown love also entails a volition and is not simply composed of intentional feelings.[63] Moreover, all axiological acts of value-intending and all conative acts of willing are founded on some doxic grounds or other.[64] In this perspective, love is a complex combination of axiological, conative, and doxic acts. Despite this intentional richness, love is an emotive experience *par excellence*, since a set of axiological acts—not conative or doxic ones—governs its manner of directedness and regard. Moreover, unlike other positive emotions and feeling-states, such as respect and joy for example, love discloses positive values with infinite horizons of endless and endlessly widening possibilities of realization, correlative to expanding communities of valuing subjects.

Husserl's threefold intentional analysis secures his ethics against traditional philosophical dichotomies, most importantly, against the dichotomies between intellect and emotion, and judgment and feeling. Instead of theorizing faculties of two monsters—a heartless head and a headless heart—he puts the whole person at the center of his mature ethics. Thus, his discussion of love does not operate by any contrastive distinctions between reason and emotion, or love and hatred,[65] but is informed by his reflective inquiries into concrete human persons, with heads, hearts, and hands and whole bodies that operate in the world. At the same time, Husserl expands human-centered accounts by conducting eidetic inquiries into non-human variants

[62] E.g., Hua42, 401; Hua43/2, 508; cf. Heinämaa 2020; 2023. Some scholars contend that love, in Husserl's analysis, is neither simply an emotive act nor an act of the will (e.g., Sowa and Vongehr 2013, cii, note 4). John Drummond, for example, distinguishes between four different types of emotive units—emotions, affective traits, character traits, and sentiments—and argues that love actually is not an emotion but a comprehensive sentiment that can be manifested by several different emotions, e.g., affection, admiration, sympathy, pride, grief, fear, and jealousy (Drummond 2020b). My interpretation differs in proposing that love indeed is an emotion for Husserl but an emotion which entails, in addition to deeply personal axiological acts of valuing, also conative acts that allow the loving subject to commit herself to her love-object in an open future, to its wellbeing and development. The conative acts involved in loving are, however, founded on the axiological acts that establish the relation in the first place.

[63] I have argued elsewhere that it is this conative structure, the structure of commitment, that opens genuine love—both in case of interpersonal relations and in case of vocational undertakings—onto an indetermined future and allows love to expand and grow in its objective coverage via transitive relations of valuing (Heinämaa 2020; 2023).

[64] The doxic foundations of axiological and conative acts must not be conceptualized as any kind of substrates that merely would bear or carry the additive senses—value and purpose—posited by axiological and conative acts, without themselves being affected in any way by them. The function of axiological and conative acts is not additive or supplementary, as if these acts would merely attach extra or external features to ready-made objects. Instead of supplying coatings or garments to doxic objectivities, axiological and practical acts reformat and restructure these objectivities themselves. Their operations are thus much more profound and comprehensive than operations of any additive acts; they affect the given objectivities and change them in characteristic manners. To use the widely applied metaphor of clothing, we can say that axiological acts do *not* clothe doxic objects with value-features, but provide given objects with new internal articulations and novel relations to other objectivities (Hua37, 180–181; cf. Jardine 2020).

[65] For a phenomenological critique of the love-hate contrast, see Kolnai 1998; cf. Steinbock 2021.

of persons. Such subjects may drastically differ from us and in several respects, but they are still equal to us in needing some kind of embodiment to perform acts of caring analogous to our human feeling and doing.[66] As pointed out above, Husserl's broadened concept of personhood includes personalities of higher order. These are organizations and institutions of individuals that are motivated by shared emotions, guided by common decisions, and operative in joint actions. By thus reworking the concepts of personhood, Husserl is able to develop his individual ethics into a communal one. This is the context in which his characterizations of philosophers as "functionaries of humankind" need to be interpreted.[67]

3. Values of Love and the Imperative of Loving

Value of love or *love-value* (*Liebeswert*) is the central concept of Husserl's late ethics. By this concept, Husserl characterizes a particular category of values that differ in several crucial respects from both objectifiable values and from the moral values that regulate interhuman exchanges and interactions. Values of love are personal in the fundamental sense that the acts that constitute their meanings emerge from the deepest cores of persons.[68] As such they are able to regulate not just individual actions or action types, but personal lives as comprehensive wholes.[69]

Husserl's category of love-values entails values that we attach to persons, both individual and collective, but they are not reducible to these. It also entails the aesthetic value of beauty and the epistemic value of truth as well as the values of the activities and practices that realize such idealities or promote their realization. Thus, values of love are "personal" merely in the sense that they draw their axiological sense from the deepest cores of valuing persons and not in the sense that they would necessarily be targeted at persons or established between persons.[70] Therefore, the ethics that is built on love-values, in Husserl's sense, is not focused on human-human relations but also covers our relations to beings of other sorts (God, nature, living beings, arts, sciences…) and, most fundamentally, it urges us to reform our relations with ourselves.[71]

As pointed out, Husserl's threefold analysis shows that loving is not composed of one axiological act, neither a valueception nor a value-judgment, but has a more complex structure: an unbound stream of feeling and value-intending issues from the deepest axiological core of the self and in such a way that it commits the ego to the beloved, its possibilities and its future. Thus, a conative element is activated but

[66] Cf. Miettinen 2013; Carr 2017.

[67] Hua6; Hua27; Hua29; cf. Miettinen 2020.

[68] HuaMat9, 146 note 1; 355; cf. Sowa and Vongher 2013, civ.

[69] Crespo 2018; Heinämaa 2020; 2023.

[70] Heinämaa 2020.

[71] Hua27, 33, 39; Hua42, 305–306.

one that is informed by the primary animating feeling. What is at issue in this manner of intending is a deep and deeply egoic power and, moreover, one that individuates the ego, distinguishes the ego as a specific kind of lover from all other loving selves. In one manuscript, Husserl writes: "I am who I am, and the individual specificity shows itself in that I love, as the one who I am, exactly as I love, that exactly this [one] calls me and that [one] does not."[72]

Thus understood, values of love are not quantifiable or comparable with other values as objectively given values are, and are not insertable into objective value hierarchies or reducible to elements of such hierarchies. They oblige the valuing person absolutely and unconditionally, and not only on the condition that they are isolated from some other values or on the condition that other values are suspended.

Another crucial implication of the deeply egoic sources of full-grown loving is that we all are dependent on one another in our attempts of navigating the broad field of values. Because each of us can relate in original love only to particular persons, specific activities, and/or certain idealities, and not to everything that is beautiful and good, we need one another to identify what can be loved and what deserves loving and thus to broaden our spectrum of love-values.[73]

The obligation of love is to care for the loved one, for the realization of its dormant dimensions[74] and for the actualization of its potentiality,[75] and to promote the perfection of such realizations and actualizations. This concerns equally realizations by ourselves and possible realizations by others. When we love art, for example, in this deeply personal manner, then we work to further and support artistic and aesthetic practices and their results, but at the same time also appreciate and cherish and care for the contributions by and efforts of all other art-lovers. Such co-lovers can include, in the case of art, not just artists but also art critics, art historians, and art-loving audiences, contemporary and past as well as future.[76]

This implies that Husserl's concept of personally established love-values, that is, values that stem from personal depths, is broader than the traditional category of moral values. It entails much more than the altruistic values of generosity and kindness or the equalitarian ones of harmony, equality, respect, and tolerance; most importantly, also epistemic and aesthetic values as well as the uniquely individual values characteristic of singular persons.[77] As such the category of love-values is morally neutral. However, all personal values of love can, in Husserl's account, be renewed and re-established as ethical values when subjected to radical self-critique and the demand of evidencing.[78] This means that both artists and scientists can prac-

---

[72] Ms. B I 21/60a, quoted in Sowa and Vongehr 2013, cix–cx.

[73] Hua42, 201.

[74] E.g., Hua27, 121.

[75] When we love a person, then the intentional object of our love is not a ready-made being but a manner of becoming (Hua5, 15–17; Hua42, 235–236, 387). Thus, our obligation as lovers in such cases is to care for the actualization and development of a potentiality (Hua42, 383–384).

[76] Hua42, 390–391; cf. Heinämaa 2023.

[77] Hua42, 278–279, cf. 343.

[78] Hua27, 29–30; cf. 42; Hua37, 238; Hua42, 35, 353–354; cf. Rinofner-Kreidl 2022.

tice their professions either in naïve forgetfulness or ethically, by radically renewing their vocational valuations.

> So art is a "vocation" for the genuine artist, and science a "vocation" for the true scientist (the "philosopher"); it is a field of spiritual activities and accomplishments, for which s/he knows that s/he "has a calling," so that only the pursuit of these goods can give her the most "inner" and most "pure" satisfaction, and each succeeding can give her the consciousness of "happiness" [*Seligkeit*]. (…) The true artist (...) as such is not yet a genuine[79] human being in the highest sense. A genuine human being can, however, be a true artist, but *can* be such only *if* ethical self-regulation *demands* this from him (…).[80]

Personal values, and only such values, ground human vocations and organize our lives into vocational wholes. The ethical imperatives that Husserl spells out in his Kaizo essays and his late lectures on ethics in 1920 and 1924 concern such wholes, already structured by personal values of love and organized into vocational projects and sets of tasks. This means that an ethical way of living can be established only on the basis of a self-regulating vocational life and cannot be composed of individually regulated actions in abstraction.

> My whole life lies in front of me, and in front of me lies the surrounding world which extends itself around me. What I can accomplish there operates as the basis for my deliberation, and the best that I can accomplish there now and in the future is my obligation [*Gesolltes*], the obligation of an individual.[81]

Husserl's explorative investigations of love-values influenced his understanding of the content and function of the categorical imperative. In his early lectures he had further developed the anti-Kantian formulations of the imperative that he found in Brentano's ethics and that command the subject always to will unconditionally the best possible among the achievable goods: "Do the best among the goods that are reachable in your current practical sphere as a whole."[82] Husserl recapitulates such Brentanian formulations in his *Kaizo* essays and later lectures on ethics, but gives them a new personalistic, situational, and holistic twist:

[79] "Genuine" here is a translation of the German word "echt" which can also be translated by "true." Husserl's demand for the ethical renewal of all *particular* vocations can itself be formulated in vocational terms. This is because, in addition to particular vocations, he also discusses a common vocational call which is addressed to all human subjects equally as rational agents and which impels them to be and become authentic (*eigentlich*) human beings (e.g., Hua42, 250–252, 383–389; cf. Drummond 1995). By "authenticity" in such contexts, he means rationality as evidencing. This implies that our common human call is to renew all our particular vocations in a manner that allows us to live in full evidence.

[80] Hua27, 28–29.

[81] Hua37, 252–253; cf. Hua27, 37, 40–48; Hua42, 239–240, 305, 390, 502.

[82] Hua28, 137. Cf. "Do the best among the goods that are reachable in your current practical sphere as a whole" (Hua28, 142). Formulations such as this, found both in Brentano and in Husserl, can be interpreted as forms of ideal consequentialism (I am grateful to John Drummond for pointing out this implication). Cf. Trincia 2007; Rinofner-Kreidl 2010; Cabrera 2017; Heffernan 2022; Heinämaa 2025.

> Do from now on and without hesitation always the best, your best, grasp it in norm-cognition and will it in norm-conscious volition.[83]

> What I must is determined by the 'I can,' and what I can differs from what any other can.[84]

However, Husserl had serious doubts about the tenability of the Brentanian reformulation of the imperative already when he gave his early lectures course on ethics at the beginning of the century. In the 1919–1920 lecture course, *Introduction to Philosophy* (*Einführung in die Philosophie*), these doubts had grown into a systematic suspicion and critique which led him to conclude that the Brentanian imperative has to be either abandoned or seriously delimited.[85] At the end, Husserl was ready to conclude that Brentano's notion of *the best achievable practical good* misses the core meaning of the ethical ought and makes sense merely in specific—profitable and fortunate—circumstances and is therefore insufficient for universal ethics.

Ultimately Husserl's reflections on love led him to abandon his early Brentanian starting points. His early suspicion grew into a straightforward critique, explicated in his lecture notes and manuscripts from the 1920s as follows:

> Thus the Brentanian rule is insufficient. Each [human being] has her or his absolute ought, and her or his choice is carried out in the question "What ought I?," and when I ought to do several [then the question is] "Which is now my necessary," not simply "Which is better in the comparison of goods?" Playing a sonata by Mozart is more beautiful than washing the child but the latter is duty, since now is its time. (…) The whole ethics of the highest practical good, as developed by Brentano and which I accepted in its essential features, cannot be the last word. It requires basic limitations! It does not do justice to vocation and inner calling.[86]

Husserl backs up this insight by discussing agentive cases in which the Brentanian imperative turns out to be inadequate or futile. For genuine ethics, he claims, it is not sufficient to study habitual ordinary cases or cases that occur in profitable and fortunate circumstances. Instead, ethics as rigorous reflection demands that we also cover limit cases.[87]

---

[83] Hua37, 253. On the situational and contextual aspects of Husserl's formulations, see Drummond 1995.

[84] Hua37, 252–253; cf. Hua27, 37, 40–48; Hua42, 305, 390. Bernard Obsieger importantly points out that this holistic and existential mode of inquiry is indebted to Husserl's reading of Kierkegaard (Obsieger 2022; cf. Hua43, 234, 253–257, 285).

[85] HuaMat9, 132. Husserl credits his student and colleague Moritz Geiger for alarming him about the core problem of Brentanian ethics already in 1907: "It is obvious that an ethics which is carried out according to the mere categorical imperative like it was, following Brentano, taken as the basis here, is not ethics at all. I have already reverted back to my old lines of thought, although Geiger already made the justified objection to me in 1907 that it would be ridiculous to demand of a mother to deliberate first, if the fostering of her child would be the best [thing to do] within her practical domain" (Hua28, 419–422; Melle 1988, xlvi–xlvii; cf. Sowa and Vongehr 2013, c–civ; Averchi 2015). On Husserl's indebtedness to Geiger, see Drummond 2018; Crespo 2021.

[86] Hua42, 390–392.

[87] For a more complete explication of this argument and a treatment of some possible counterarguments, see Heinämaa 2025.

For this purpose, Husserl asks us first to imagine a situation in which all prospects of accomplishing anything good, or better or best, are lost and lost for everyone. To be sure, such a hopeless world is far from our normal agentive setting, but it is not an impossibility nor completely foreign to the collective experiences of the 20th century, in the wake of Somme and Verdun, Dachau and Treblinka, the Gulags, Hiroshima and Nagasaki, Rwanda, Guantanamo, Xinjiang...

Husserl then imaginatively inserts two different kinds of agents in this dystopic, apocalyptic situation in which all prospects of achieving anything good are lost. One agent is a "performance machine" or "achievement apparatus" (*Leistungsmachine*) who unhesitatingly follows the Brentanian imperative of always doing and aiming to do the good and the best practically possible, the best that he can. The other agent is a parent, a mother, who knows the Brentanian imperative but, in addition to it, is able to follow a personal obligation of loving and lovingly caring for her loved one.[88] Husserl's argument is that in such an extreme situation, the performance machine would be paralyzed by futile comparisons, since none of its possible actions would present itself as better than the others and no best one among them could be identified. The mother, in contrast, would act immediately since no change in situational parameters changes her loving obligation:

> [It] may be and may be certain [that the world is destined to destruction and the human kind 'develops' without progress, without becoming better], but a thousand times more certain is that I must not abandon my child, that I should lovingly care for the child; also that the child's bodily and spiritual becoming good and thereby the child's being good is an absolute value which I should and ought to set as a goal for myself. If I live according to this, then I am good so far and [do] my duty. Moreover, that is and remains good, whether the world still has sense or not.[89]

Husserl then asks himself: "[I]f the mother knew for sure that tomorrow, that in one hour, the end of the world would come, would she as a true mother fail during this one hour to give her child the loving care, consolation. etc.?" and then contends: "[I]s it not similar with everything ethical?"

The mother is not a model of ethical agency or paradigm of moral subjectivity for Husserl but an imaginary example that allows us to pay heed to a crucial difference of sense in the certainties, characteristic of the two imperatives, the imperative to do the best possible and the commandment of love. Whereas the former is absolute in its form merely, due to its lack of material specification, the second one is absolute in its verbal specificity and reference to the intentional activity of loving, with personal axiological and conative sources. Whereas the first one commands us to compare the goodness of our goals in terms of their possibilities of realization, the second one commands us to act lovingly, whatever comes.[90]

We can summarize Husserl's revision of the Kantian imperative by saying that he started with Brentano's reformulations that regulate doing, willing, and acting, but

---

[88] Hua42, 309–310, cf. 241, 391–392; cf. Heinämaa 2020.

[89] Hua42, 310.

[90] Hua42, 310–311.

enriched these rules by imperatives of being good and becoming one's best person; however, when his reflections on love proceeded, he realized the fundamental ethical significance of the commandment of love. His investigations thus move from conative imperatives of doing and willing (*Tunsollen*) to doxic imperatives of being and becoming (*Seinsollen*), and ultimately to an axiological imperative of emotional life, "Love!."[91]

Common sense thinking proposes that it is senseless to try to command emotions or order human subjects to have emotions and experience feelings, let alone dictate love. This is because emotions as feelings are taken to be merely passive affective and receptive states without any active engagement or agency on the part of the emotive subject. Against such common sense and popular scientific teachings, Husserl contends that in addition to their governing axiological parts, most emotions, if not all, also entail conative moments. In the case of genuine love, the conative part, activated and guided by the axiological one(s), is rooted in the deepest core of the loving persons. Thus, an imperative of loving is not only meaningful but also necessary: deeply personal love essentially entails a conative act founded on the axiological one by which the subject commits herself and dedicates her life to the promotion of the valued object, its perfection in beauty and goodness, and its development and enrichment in further disclosing activities.

Moreover, personal values of love tie us to one another as axiological subjects, capable of valuing in absolute, unconditional manners. On the one hand, we are bound by mutual relations of care to all others who personally value what we value, be it a friend or a child, a country or a nation, a field of the arts or one of the sciences. On the other hand, we are also bound by relations of care to any person who is able to love something in the *structurally similar* personal and personally deep manner as we ourselves are.[92] We may not love the same objects that our fellow subjects love, but insofar as we ourselves are capable of valuing something deeply, from the deepest cores of our intentional lives, we can identify similar valuations in others and relate to them in empathy, sympathy, and loving care:

> Of course, the fulfilment of my personal absolute ought is an originary lived experience only for me and as such preferred [*bevorzugt*]. But the law of love reigns here. As soon as I have the other given as the subject of her personal ought in empathy (...) I "have to" sympathize with her, love her; and to encourage or promote her is my absolute ought.[93]

Our capacity of feeling and valuing in this deeply personal manner binds us together into a *community of love* with a potential to expand in larger and larger circles.[94] This community is not just a community of feeling subjects, since love as an emotion entails both axiological and conative factors. Rather than being simply

---

[91] Hua42, 397; cf. Crespo 2015; Heffernan 2022.

[92] Hua42, 355; cf. Heinämaa 2020.

[93] Ms. F I 24, 38a, b, cited in Loidolt 2012, 25.

[94] Hua42, 354–355; 390–391, 462, 513–514; cf. Buckley 1996; Melle 1991; 2002; Heinämaa 2020; Crespo 2022.

connected by affects, feelings, and drives, the members of a community of love are actively committed to one another both in valuing and in value-informed willing.[95]

In Husserl's analysis, the most fundamental value of all is the value that characterizes persons who all are capable of loving and lovingly committing themselves.[96] This is because only such beings are able to disclose and realize absolute goods and beauties in the world common to all.[97]

Familiarity with Husserl's explorative investigations into love thus helps us understand the parallel development of his phenomenological theory of persons and philosophical ethics. Moreover, it sheds light on emotive and axiological aspects of his social ontology and social phenomenology, and thereby allows us to rethink what he meant in the 1930s by referring to philosophers as "the functionaries of mankind."

By familiarizing us with Husserl's concept of love, we can also discover philosophically crucial and fruitful relations between classical phenomenology and other philosophical approaches. The most important parallels include alternative phenomenological approaches, developed by Husserl's realist and existentialist colleagues as well as contemporary phenomenologists, but also analytically orientated discussions on personhood, agency, and emotions.[98] Also classical philosophical discussions of emotions provide illuminative points of comparison. Husserl's approach resonates most importantly with those of Plato, Aristotle, and the Stoic thinkers as well as Descartes, Leibniz, and Kierkegaard.

Finally, Husserl's concepts of higher order personalities and the community of love shed light on several phenomena under the headings "collective intentionality" and "group-formation," vividly discussed in contemporary social ontology and social phenomenology. At the same time, we find new resources for a critical encounter with neo-Kantian, neo-Hegelian, and neo-Aristotelian approaches, dominant in contemporary political and moral philosophy. In this respect, Husserl's most significant result concerns the constitution of personal identities, both individual and collective. His explorative thinking locates the concrete sources of all identities in vocational commitments, and ultimately in our capacities of empathizing in emotion, valuing, and willing. It thus challenges dominant theories according to which human identities, both individual and social, are grounded in relations of conflict, recognition, projection, and/or imitation.

---

[95] Caminada 2016; 2019.

[96] E.g., Hua42, 316, cf. 256–257, 287–288, 304.

[97] This argument is based on two insights: (i) Husserl understanding of the *essence* of person as an open and dynamic whole, mentioned above, and (ii) his original conceptualization of the value of person as *καλοκαγαθία* (*kalokagathia*), that is, as a peculiar combination of *good-values* and *beauty-values* (*Gutwert*, *Schönheitswert*) and such that the being of a person is valued purely *for the sake of* (*umwillen*) her or his beautiful manner of appearing (Hua43/2, 248–258; cf. Hua37, 189).

[98] E.g., Goldie 2012; 2002; Helm 2001; Frankfurt 2004; 1998; Williams 1981.

Bibliography

Averchi, Michaele 2015: "Husserl and Geiger on feelings and intentionality," in *Feeling and Value, Willing and Action*, eds. Marta Ubiali and Maren Wehrle, Cham: Springer, 105–118.

Benois, Jocelyn 2002: "Non-objectifying acts," *One Hundred Years of Phenomenology: Husserl's* Logical Investigations *Revisited*, eds. Dan Zahavi and Frederick Stjernfelt, Dordrecht: Springer, 41–49.

Bernet, Rudolf 1994: "An intentionality without subject and object," *Man and World*, vol. 27, no. 3, 231–255.

Bernet, Rudolf 2002: "Different concepts of logic and their relation to subjectivity," in *One Hundred Years of Phenomenology: Husserl's* Logical Investigations *Revisited*, eds. Dan Zahavi and Fredrik Stjernfelt, Dordrecht: Springer, 19–29.

Brentano, Franz [1889] 1969: *Vom Ursprung sittlicher Erkenntnis*, ed. Oskar Kraus, Hamburg: Meiner. In English *The Origin of the Knowledge of Right and Wrong*, London: Routledge, 2010.

Brentano, Franz [1952] 1954: *Grundlegung und Aufbau der Ethik*, ed. Franziska Mayer-Hillebrand, Bern: Franke.

Breuer, Irene 2019: "The ego as moral person: Husserl's concept of personhood in the context of his later ethics," *Miscellanea Anthropologica et Sociologica*, vol. 20, no. 1, 15–35.

Brink, David O. 2019: "Normative perfectionism and the Kantian tradition," *Philosopher's Imprint*, vol. 19, no. 45, 1–28.

Buchanan, Robert Agnus: "History of technology," *Encyclopedia Britannica*, https://www.britannica.com/technology/history-of-technology

Buckley, Philip R. 1996: "Husserl's rational '*Liebesgemeinschaft*'," *Research in Phenomenology*, vol. 26, no. 1, 116–129.

Buckley, Philip R. 2000: "Personality of higher order: Husserlian reflections on the Québec problem," in *Phenomenology of the Political*, eds. Kevin Thompson and Lester Embree, Dordrecht: Springer, 105–120.

Cabrera, Celia 2017: "La apropiración husserliana del imperativo categórico," *Arte*, vol. 29, no. 1, 29–58.

Caminada, Emanuele 2016: "Higher order persons: An ontological challenge?," *Phenomenology and Mind*, no. 1 (2016), 152–157.

Caminada, Emanuele 2019: *Vom Gemeingeist zum Habitus: Husserls Ideen II: Sozialphilosophische Implikationen der Phänomenologie*, Dordrecht: Springer.

Carr, David 1987: "Personalities of higher order," in *Interpreting Husserl: Critical and Comparative Studies*, Dordrecth: Martinus Nijhoff, 267–279.

Carr, David 2017: "'...so etwas wie Leiblichkeit': On social embodiment," *Yearbook for Eastern and Western Philosophy*, no. 2, 91–103.

Carta, Emanuela 2021: "Husserl on eidetic norms," *Husserl Studies*, vol. 37, no. 2, 127–146, https://link.springer.com/article/10.1007/s10743-020-09284-5

Cavallaro, Marco and George Heffernan 2019: "From happiness to blessedness: Husserl on *eudaimonia*, virtue, and the best life," *Horizon*, vol. 8, no. 2, 353–388.

Cavallaro, Marco and George Heffernan (eds.) 2022: *The Existential Husserl*, Dordrecht: Springer, 251–275.

Crespo, Mariano 2015: "Husserl on personal aspects of moral normativity," *Ethical Perspectives*, vol. 22, no. 4, 699–722.

Crespo, Mariano 2018: "From empathy to sympathy: On the importance of love in the experience of other," in *Husserl's Phenomenology of Intersubjectivity: Historical Interpretations and Contemporary Applications*, eds. Frode Kosjavik, Christian Beyer, and Christel Fricke, London: Routledge, 235–248.

Crespo, Mariano 2021: "The ethics of Husserl and his contemporaries (Lipps, Fender, and Geiger)," in *The Husserlian Mind*, ed. Hanne Jacobs, London: Routledge, 367–376.

Crespo, Mariano 2022: "Communalization (*Vergemeinschaftung*) through ethical love: A phenomenological account," in *Ways of Being Bound: Perspectives from Post-Kantian Philosophy and*

Relational Sociology, eds. Particio A. Fernández, Alejandro Néstor García Martínez and José M. Torralba, Dordrecht: Springer, 105–116.

De Monticelli, Roberta 2021: *Towards a Phenomenological Axiology: Discovering What Matters*, Cham: Palgrave/Macmillan.

Donohoe, Janet 2004: *Husserl on Ethics and Intersubjectivity: From Static to Genetic Phenomenology*, Toronto, Buffalo, London: University of Toronto Press.

Drummond, John 1995: "Moral objectivity: Husserl's sentiments of the understanding," in *Husserl Studies*, vol. 12, no 2, 165–183.

Drummond, John 2003: "The structure of intentionality," in *The New Husserl: A Critical Reader*, ed. Donn Welton, Indianapolis: Indiana University Press, 65–92.

Drummond, John 2006: "Respect as a moral emotion: A phenomenological approach," *Husserl Studies*, vol. 22, no. 1, 1–27.

Drummond, John 2009: "Feelings, emotions, and truly perceiving the valuable," *Modern Schoolman*, vol. 86, no. 3–4, 363–379.

Drummond, John 2013: "The intentional structure of emotions," *History of Philosophy & Logical Analysis*, vol. 16, no. 1, 244–263.

Drummond, John 2015a: "Exceptional love?," in *Feeling and Value, Willing and Action: Essays in the Context of Phenomenological Psychology*, eds. Marta Ubiali and Maren Wehrle, Cham: Springer, 51–69.

Drummond, John 2015b: "Husserl's phenomenological axiology and Aristotelian virtue ethics," in *New Perspectives on Aristotelianism and its Critics*, eds. Miira Tuominen, Sara Heinämaa and Virpi Mäkinen, Leiden, Boston: Brill, 179–195.

Drummond, John 2016: "Time and the 'antinomies' of deliberation," in *Time and the Philosophy of Action*, eds. Roman Altshuer and Michael J. Sigrist, London: Routledge.

Drummond, John 2018: "Husserl's middle period and the development of his ethics," in *The Oxford Handbook of the History of Phenomenology*, eds. Dan Zahavi, Oxford: Oxford University Press, 135–154.

Drummond, John 2019: "Intentionality and (moral) normativity," *Normativity, Meaning, and the Promise of Phenomenology*, eds. Matthew Bruch, Jack Marsh and Irene McMullin, London: Routledge, 101–119.

Drummond, John 2020a: "Political community," in *Phenomenology of the Political*, eds. Kevin Thompson and Lester Embree, Dordrecht: Springer, 29–53.

Drummond, John 2020b: "The varieties of affective experience," in *The Routledge Handbook of the Phenomenology of the Emotions*, eds. T. Szanto and H. Landweer, London: Routledge, 239–249.

Drummond, John 2021a: "Husserl's complex concept of objectivity," in *The Husserlian Mind*, ed. Hanne Jacobs, 327–339.

Drummond, John 2021b: "Phenomenological method and contemporary ethics," *Continental Philosophy Review*, vol. 54, no. 3, 123–138.

Drummond, John 2023: "Community: A unified disunity?," *Continental Philosophy Review*, vol. 56, no. 3, 401–417.

Drummond, John and Sonja Rinofner-Kreidl 2020: "Morality and emotions," in *The Routledge Handbook of Phenomenology of Emotion*, eds. Thomas Szanto and Hilge Landweer, New York: Routledge, 288–298

Drummond, John and Sonja Rinofner-Kreidl 2018: "Introduction," in *Emotional Experiences: Ethical and Social Significance*, London, New York: Rowman & Littlefield, 1–13.

Enquist Källgren, Karolina and Íngrid Vendrell Ferran 2022: "Scheler and Zambrano: On a transformation of the heart in Spanish philosophy," *History of European Ideas*, vol. 48, no. 5, 634–649.

Ferrarello, Susi 2016: *Husserl's Ethics and Practical Intentionality*, London, New York: Bloomsbury.

Ferrarello, Susi 2023: "Husserl and Stoicism: The theory of meaning as a bridge to mindfulness," *Journal of Humanistic Psychology*, vol. 26, no. 6, 1–19.

Fisette, Denis 2021: "Emotions and moods in Husserl's phenomenology," in *The Husserlian Mind, The Husserlian Mind*, ed. Hanne Jacobs, London: Routledge, 220–231.
Flynn, Molly Bridgid 2010: "Philosophy and the integrity of the person: The phenomenology of Robert Sokolowski," *The Reception of Phenomenology in North America*, ed. Michaela Beatrice Ferri, Dordrecht: Springer, 187–204.
Flynn, Molly Bridgid 2012: "The cultural community: An Husserlian approach and reproach," *Husserl Studies*, vol. 28, no. 1, 25–47.
Goldie, Peter 2000: *The Emotions: A Philosophical Exploration*, Oxford: Oxford University Press.
Goldie, Peter 2012: *The Mess Inside: Narrative, Emotion, and the Mind*, Oxford: Oxford University Press.
Gruyer, Paul 2011: "Kant and virtue ethics," *Perfecting Virtue: New Essays on Kantian Ethics and Virtue Ethics*, eds. Lawrence Jost and Julian Wuerth, Cambridge: Cambridge University Press.
Gruyer, Paul 2016: "Kantian perfectionism," in *The Virtues of Freedom: Selected Essays on Kant*, De Guryter, 70–86.
Gschwandtner, Christina M. 2022: "Guarding the heart: The phenomenality of the heart in early Christian asceticism," in *Phenomenology and Perspectives on the Heart*, ed. Anthony J. Steinbock, Dordrecht: Springer, 19–50.
Gubser, Michael D. 2014: "The blueprint of a new heart: Max Scheler and the order of love," *The Far Reaches: Phenomenology, Ethics, and Social Renewal in Central Europe*, Stanford, California: Stanford University Press, 80–100.
Frankfurt, Harry G. 1998: *The Importance of What We Care About*, Cambridge: Cambridge University Press.
Frankfurt, Harry G. 2004: *The Reasons of Love*, Princeton and Oxford: Princeton University Press.
Hahn, Collin J. 2012: *The Concept of Personhood in the Phenomenology of Edmund Husserl*, Doctoral Dissertation, Marquette University.
Hart, James G. 1992: *The Person and the Common Life: Studies in Husserlian Social Ethics*, Dordrecht: Kluwer Academic Publishers.
Hart, James G. 2006: "The absolute ought and the unique subject," *Husserl Studies*, vol. 22, no. 3, 223–240.
Hart, James G. 2009: *Who One Is: A Transcendental-Existential Phenomenology; Book 1: A Meontology of the "I": A Transcendental Phenomenology; Book 2: Existenz and Transcendental Phenomenology*, Dordrecht: Springer.
Hart, James G. 2021: "Transcendental-phenomenological ontology of persons and the singularity of love," *Eidos: A Journal for Philosophy and Culture*, vol. 4, no. 4, 136–174.
Hartimo, Mirja 2018: "Radical *Besinnung* in *Formale und transzendentale Logik* (1929)," *Husserl Studies*, vol. 34, no. 3, 247–266.
Hartimo, Mirja 2020: "Husserl on 'Besinnung' and formal ontology," in *Metametaphysics and the Sciences: Historical and Philosophical Perspectives*, eds. Frode Kjosavik and Camilla Sereck-Hanssen, New York: Routledge, 200–215.
Heffernan, George 2022: "The development of Husserl's categorical imperative: From universal ethical legislation to individual existential exhortation," in *The Existential Husserl*, eds. Marco Cavallaro and George Heffernan, Dordrecht: Springer, 251–275.
Heinämaa, Sara 2007: "Selfhood, consciousness, and embodiment: A Husserlian approach," in *Consciousness: From Perception to Reflection in the History of Philosophy*, eds. Sara Heinämaa, Pauliina Remes and Vili Lähteenmäki, Dordrecht: Springer, 311–328.
Heinämaa, Sara 2014: "Husserl's ethics of renewal: A personalistic approach," in *New Perspectives to Aristotelianism and Its Critics*, eds. Miira Tuominen, Sara Heinämaa and Virpi Mäkinen, Leiden, Boston: Brill, 196–212.
Heinämaa, Sara 2020: "Values of love: Two forms of infinity characteristic of human persons," *Phenomenology and the Cognitive Sciences*, vol. 19, no. 3, 431–450, https://doi.org/10.1007/s11097-019-09653-2
Heinämaa, Sara 2021: "*Cartesian Meditations*: Husserl's pluralistic egology," *The Husserlian Mind*, ed. Hanne Jacobs, London: Routledge, 38–49.

Heinämaa, Sara 2022a: "Self – A phenomenological account: Temporality, finitude and intersubjectivity," in *Empathy, Intersubjectivity, and the Social World: The Continued Relevance of Phenomenology, Essays in Honour of Dermot Moran*, eds. Anna Bortolan and Elisa Magri, Berlin: De Gruyter, 187–200.

Heinämaa, Sara 2022b: "Varieties of normativity: Norms, goals, values," in *Contemporary Phenomenologies of Normativity: Norms, Goals, Values*, eds. Sara Heinämaa, Mirja Hartimo and Ilpo Hirvonen, London, New York: Routledge, 19–42.

Heinämaa, Sara 2023: "Personal, communal and temporal structures of vocational life: A Husserlian account," *Continental Philosophy Review*, eds. Dan Zahavi, Thomas Szanto and Patricia Meindl, 461–481.

Heinämaa, Sara 2024: "'Gemüt' as 'emotion' vs. 'heart': Systemic, conceptual and terminological arguments," unpublished manuscript.

Heinämaa, Sara 2025: "Personal values and the commandment of love: Husserl's reformulation of the categorical imperative," *Essays on Husserl's Kaizo Articles: Historical Context, Systematic Questions, Global Perspectives*, eds. Genki Uemura, Emanuele Caminada and Julia Jansen, Alber, forthcoming.

Helm, Bennet W. 2001: *Emotional Reason: Deliberation, Motivation, and the Nature of Value*, Cambridge: Cambridge University Press.

von Hildebrand, Dietrich [1965] 1970: *The Heart: An Analysis of Divine and Human Affectivity*, ed. John Henry Crosby, South Bend, Indiana: St. Augustine's Press.

Husserl, Edmund Hua1: *Cartesianische Meditationen und pariser Vorträge*, *Husserliana I*, ed. Stephan Strasser, The Hague: Martinus Nijhoff, 1950. In English: *Cartesian Meditations*, trans. Dorion Cairns. The Hague: Martinus Nijhoff, 1960.

Husserl, Edmund Hua3: *Ideen zu einer reinen Phänomenologie und phänomenologischen Philosophie, Erstes Buch: Allgemeine Einführung in die reine Phänomenologie*, *Husserliana II*, ed. Walter Biemel, The Hague: Martinus Nijhoff, 1950. In English *Ideas Pertaining to a Pure Phenomenology and to a Phenomenological Philosophy, First Book: General Introduction to a Pure Phenomenology*, trans. Fred Kersten, The Hague, Boston, Lancaster: Martinus Nijhoff, 1983.

Husserl, Edmund Hua3/1: *Ideen zu einer reinen Phänomenologie und phänomenologischen Philosophie, Erstes Buch: Allgemeine Einführung in die reine Phänomenologie*, *Husserliana III/1*, ed. Karl Schumann, The Hague: Martinus Nijhoff, 1976. In English *Ideas Pertaining to a Pure Phenomenology and to a Phenomenological Philosophy, First Book: General Introduction to a Pure Phenomenology*, trans. Fred Kersten, The Hague, Boston, Lancaster: Martinus Nijhoff, 1983.

Husserl, Edmund Hua4: *Ideen zu einer einen Phänomenologie und phänomenologischen Philosophie, Drittes Buch: Die Phänomenologie und die Fundamente der Wissenschaften*, *Husserliana V*, ed. Marly Biemel, The Hague: Martinus Nijhoff, 1952. In English: *Ideas Pertaining to a Pure Phenomenology and to a Phenomenological Philosophy, Second Book: Studies in the Phenomenological Constitution, Collected Works II*, trans. Richard Rojcewicz and André Schuwer, Dordrecht, Boston, London: Kluwer, 1993.

Husserl, Edmund Hua4/5: *Ideen zu einer reinen Phänomenologie und phänomenologischen Philosophie, Zweites Buch: Phänomenologische Untersuchungen zur Konstitution und Wissenschaftstheorie*, *Husserliana IV/V*, ed. Dirk Fonfara, The Hague: Springer, forthcoming.

Husserl, Edmund Hua6: *Die Krisis der europäischen Wissenschaften und die transzendentale Phänomenologie: Eine Einleitung in die phänomenologischen Philosophie*, *Husserliana VI*, ed. Walter Biemel. The Hague: Martinus Nijhoff, 1954. In English: *The Crisis of European Sciences and Transcendental Phenomenology: An Introduction to Phenomenological Philosophy*, trans. David Carr. Evanston: Northwestern University, 1988.

Husserl, Edmund Hua14: *Zur Phänomenologie der Intersubjektivität, Texte aus dem Nachlass, Zweiter Teil: 1921–1928*, *Husserliana XIV*, ed. Iso Kern, The Hague: Martinus Nijhoff, 1973.

Husserl, Edmund Hua15: *Zur Phänomenologie der Intersubjektivität, Texte aus dem Nachlass, Zweiter Teil: 1929–1935*, *Husserliana XV*, ed. Iso Kern, The Hague: Martinus Nijhoff, 1973.

Husserl, Edmund Hua17: *Formale und transzendentale Logik: Versuch einer Kritik der logischen Vernunft, Husserliana XVII*, ed. Paul Janssen, The Hague: Martinus Nijhoff, 1974. In English *Formal and Transcendental Logic*, trans. Dorion Cairns, The Hague: Martinus Nijhoff, 1969.
Husserl, Edmund Hua18: *Logische Untersuchungen, Erster Band: Prolegomena zur reinen Logik, Husserliana XVIII*, ed. Elmar Holenstein, The Hague: Martinus Nijhoff, 1975. In English *Logical Investigations, Volume I*, trans. J.N. Findlay, London, New York: Routledge, 2001.
Husserl, Edumund Hua19/1: *Logische Untersuchungen, Zweiter Band, Erster Teil: Untersuchungen zur Phänomenologie und Theorie der Erkenntnis, Husserliana XIX/1*, ed. Ursula Panzer, The Hague: Martinus Nijhoff, 1984. In English *Logical Investigations*, Volume 2, ed. Dermot Moran, trans. J.N. Findlay, London and New York: Routledge, 2001.
Husserl, Edmund Hua23: *Phantasie, Bildbewusstsein, Erinnerung: Zur Phänomenologie der anschaulichen Vergegenwärtigungen, Texte aus dem Nachlass (1898–1925), Husserliana XXIII*, ed. Eduard Marbach, The Hague, Boston, London: Martinus Nijhoff, 1980.
Husserl, Edmund Hua24: *Einleitung in die Logik und Erkenntnistheorie, Vorlesungen 1906/1907, Husserliana IV*, ed. Ulrich Melle, Dordrecht, Boston, Lancaster: Martinus Nijhoff, 1984.
Husserl, Edmund Hua27: *Aufsätze und Vorträge 1922–1937, Husserliana XXVII*, eds. Thomas Nenon and Hans Rainer Sepp, Dordrecht: Kluwer Academic Publishers, 1989.
Husserl, Edmund Hua28: *Vorlesungen über Ethik und Wertlehre 1908–1914, Husserliana XXVIII*, ed. Ulrich Melle, Dordrecht: Kluwer Academic Publishers, 1988.
Husserl, Edmund Hua37: *Einleitung in die Ethik, Vorlesungen Sommersemester 1920 und 1924, Husserliana XXXVII*, ed. Henning Peucker, Dordrecht: Kluwer Academic Publishers, 2004.
Husserl, Edmund Hua42: *Grenzprobleme der Phänomenologie: Analysen des Unbewusstseins und der Instinkte, Metaphysik, Späte Ethik, Texte aus dem Nachlass (1908–1937), Husserliana XLII*, eds. Rochus Sowa and Thomas Vongehr, Dordrecht: Springer, 2013.
Husserl, Edmund Hua43/2: *Studien zur Struktur des Bewusstseins, Teilband II: Gefühl und Wert, Texte aus dem Nachlass 1896–1925, Husserliana XLIII/2*, eds. Ulrich Melle and Thomas Vongehr, Dordrecht: Springer, 2020.
Husserl, Edmund Hua43/3: *Studien zur Struktur des Bewusstseins, Teilband III: Wille und Handlung, Texte aus dem Nachlass 1902–1934, Husserliana XLIII/3*, eds. Ulrich Melle and Thomas Vongehr, Dordrecht: Springer, 2020.
Husserl, Edmund HuaMat9: *Einleitung in die Philosophie, Vorlesungen 1916–1919, Husserliana Materien IX*, ed. Hanne Jacobs, Dordrecht: Springer, 2012.
Jacobs, Hanne 2010: "Toward a phenomenological account of personal identity," *Philosophy, Phenomenology, Sciences: Essays in Commemoration of Edmund Husserl*, eds. Filip Mattens, Hanne Jacobs and Carlo Ierna, Dordrect: Springer, 333–361.
Jacobs, Hanne 2020: "Husserl, the active self, and commitment" *Phenomenology and the Cognitive Sciences*, vol. 20, no. 2, 281–298.
Jansen, Julia 2018: "Imagination de-naturalized: Phantasy, the imaginary, and imaginative ontology," in *Oxford Handbook of the History of Phenomenology*, ed. Dan Zahavi, Oxford: Oxford University Press, 676–695.
Jansen, Julia 2020: "Imagination and phantasy," in *The Routledge Handbook of Phenomenology and Phenomenological Philosophy*, eds. Daniel De Santis, Burt C. Hopkins and Claudio Majolino, London: Routledge.
Jardine, James 2020: "Edmund Husserl," in *The Routledge Handbook of Phenomenology of Emotion*, eds. Thomas Szanto and Hilge Landweer, New York: Routledge, 53–62.
Jardine, James 2022: *Empathy, Embodiment, and the Person: Husserlian Investigations of Social Experience and the Self*, Dordrecht: Springer.
Jardine, James 2024/2025: "Envisioning others without pretence: Husserl and Stein on the irreducibility, complexity and value of empathy," in *The Empathetic Emotions in the History of Philosophy*, eds. David James and Keith Ansell-Pearson, Oxford: Oxford University Press, forthcoming.
Josset, Sylvain 2021: "A 'logic of the heart'? Max Scheler, reader of Pascal?," *Dix-Septième Siècle*, vol. 290, no. 1, 9–26.

Kant, Immanuel [1979] 2009: *Metaphysik der Sitten, Werkausgabe Band VIII*, ed. Wilhelm Weischedel, Suhrkamp. In English *The Metaphysics of Morals*, trans. Mary J. Gregor, Cambridge: Cambridge University Press.

Kolnai, Aurel 1998: "The standard modes of aversion: Fear, disgust and hatred," *Mind*, vol. 107, no. 427, 581–595.

Leitlande, Gita 2022: "Authentic existence: Bridging phenomenology and Stoicism," *Problemos*, vol. 101, 92–104.

Loidolt, Sophie 2012: "The 'Daimon' that speaks through love: A phenomenological ethics of the absolute ought – Investigating Husserl's unpublished texts," in *Ethics and Phenomenology* eds. M. Sanders and J. J. Wisnewski, Lanham, Boulder, New York: Lexington, 9–38.

Luft, Sebastian 2010: "Das Subjekt als moralische Person: Zur Husserls später Reflexionen bezüglich des Personenbegriffs," in *Geist – Person – Gemeinschaft, Special Issue of Studien zur Phänomenologie und practischen Philosophie*, eds. Hans-Helmuth Gander, Lore Hühn, László Tengelyi and Günter Zöller, Vienna: Ergon Verlag, 221–240. In English "The subject as moral person: On Husserl's late reflections concerning the concept of personhood," in Investigating Subjectivity: Classical and New Perspectives, eds. G.-J. van Der Heiden, K. Novotny, Inga Römer and Lázló Tengelyi, Leiden, Boston: Brill, 25–43.

Marion, Jean-Luc [1986] 1999: *Sur le prisme métaphysique de Descartes: Constitution et limites de l'onto-théo-logie dans la pensée cartésienne*, Paris. In English *On Descartes' Metaphysical Prism*, trans. Jeffrey L. Kosky, Chicago, 1999.

Marosan, Bence Peter 2022: "Levels of the absolute in Husserl," *Continental Philosophy Review*, vol. 55, no. 1, 137–158.

Mayer, Verena and Christopher Erhard 2019: "The significance of objectifying acts in Husserl's fifth investigation," in *Husserl: German Perspectives*, eds. John Drummond and Otfried Höffe, New York: Fordham University Press, 163–192.

Melle, Ulrich 1988: "Einleitung," in Edmund Husserl: *Vorlesungen über Ethik und Wertlehre 1908–1914, Husserliana XXVIII*, ed. Ulrich Melle, Dordrecht: Kluwer Academic Publishers, xiii–xlix.

Melle, Ulrich 1991: "The development of Husserl's ethics," *Études phénoménologiques*, vol. 7, nos. 13–14, 115–135.

Melle, Ulrich 2002: "Edmund Husserl: From reason to love," in *The Phenomenological Approaches to Moral Philosophy* , eds. John Drummond and Lester Embree, Dordrecht: Kluwer, 229–248.

Melle, Ulrich 2007: "Husserl's personalistic ethics," *Husserl Studies*, vol. 23, no. 1, 1–15.

Melle, Ulrich 2019: "Objectifying and nonobjectifying acts," in *Husserl: German Perspectives*, ed. John Drummond and Otfried Höffe, New York: Fordham University Press, 193–208. Original "Objektivierende und nicht-objektivierende Akte," in *Husserl-Ausgabe und Husserl-Forschung*, ed. J.C. Ijsseling, Dordrecht, Boston, London: Kluwer Academic Publishers, 35–49.

Miettinen, Timo 2013: "The body politic: Husserl and embodied community," in *The Phenomenology of Embodied Subjectivity*, eds. Dermot Moran and Rasmus Thybo Jensen, Dordrecht: Springer, 329–346.

Miettinen, Timo 2014: "Transcendental social ontology," in *Phenomenology and the Transcendental*, eds. Sara Heinämaa, Mirja Hartimo and Timo Miettinen, New York, London: Routledge, 147–171.

Miettinen, Timo 2020: *Husserl and the Idea of Europe*, Evanston, Illinois: Northwestern University Press.

Moran, Dermot 2005: *Edmund Husserl: Founder of Phenomenology*, Malen, MA: Polity.

Moran, Dermot 2009: "The phenomenology of personhood: Charles Taylor and Edmund Husserl," *Colloquim*, vol. 3, no. 1, 80–104.

Moran, Dermot 2013: "'Die verborgene Einheit intentionaler Innerlichkeit': Husserl on history, life and tradition," *Alter: Revue de phénoménologie*, vol. 21, no. 21, 117–134.

Moran, Dermot 2017: "Husserl's layered concept of human persons: Conscious and unconscious," in *Unconsciousness Between Phenomenology and Psychoanalysis*, eds. Dorothee Legrand and Dylan Trigg, Dordrect: Springer, 3–22.

Mulligan, Kevin 1998: "The spectre of inverted emotions and the space of emotions," *Acta Analytica*, vol. 12, 89–105.

Mulligan, Kevin 2004: "Husserl on the 'logics' of valuing, values and norms," in *Fenomenologia della ragion practica: L'ethica di Edmund Husserl*, eds. Beatrice Centi and Gianna Gigliotti, Bibliopolis, 177–225.

Mulligan, Kevin 2009: "On being struck by value," in *Leben mit Gefühlen*, ed. B. Merker, Paderborn: Mentis.

Mulligan, Kevin 2017: "Intentionality, reasons and motives," *Intentionality and Action*, eds. Jesús Padilla Gálvez and Margit Gaffal, De Gruyter, 9–22.

Nenon, Thomas 1991: "Willing and acting in Husserl's lectures on ethics and value theory," *Man and World*, vol. 24, no. 3, 301–309.

Nenon, Thomas 2009: "Feelings and reason in Husserl's lectures on ethics 1920–24," at *OPO III*, Chinese University of Hong Kong, 2008, and the international conference *Space of Phenomenology: On the 150th Anniversary of Edmund Husserl*, Moskow 2009.

Obsieger, Benhard 2016: "Husserls Frage: 'Kann ich mein Leben leben, ohne dass ich es wollen kann?'," in *Phänomenologische Forschungen 2015*, eds. Karl-Heinz Lembeck, Karl Mertens, and Ernst-Wolfgang Orth, Hamburg: Felix Meiner, 323–334.

Obsieger, Bernhard 2017: "Husserlian ethics," *Intentionality and Action*, eds. Jesús Padilla Gálvez and Margit Gaffal, De Gruyter, 159–172.

Obsieger, Bernhard 2022: "Kierkegaard and Husserl," in *The Existential Husserl*, eds. Marco Cavallaro and George Heffernan, Dordrecht: Springer, 227–250.

Peucker, Hennig 2004: "Einleitung des Herausgebers," in *Einleitung in die Ethik, Vorlesungen Sommersemester 1920 und 1924*, ed. Henning Peucker, Dordrecht: Kluwer Academic Publishers, 2004, xiii–xlv.

Peucker, Hennig 2008: "From logic to the person: An introduction to Husserl's ethics," *The Review of Metaphysics, A Philosophical Quarterly*, vol. 62, no. 2, 307–325.

Ricœur, Paul 1994: *Lectures, tome 3: aux frontières de la philosophie*, Paris: Seuil.

Rinofner-Kreidl, Sonja 2010: "Husserl's categorical imperative and his related critique of Kant," in *Epistemology, Archeology, Ethics: Current Investigations of Husserl's Corpus*, eds. Pol Vandevelde and Sebastian Luft, London, New York: Continuum, 188–210.

Rinofner-Kreidl, Sonja 2011: "Handlungstheoretische und ethische Aspekte der Verantwortung," in *Ethik in Forschung und Technik: Annäherungen*, eds. Leopold Neuhold and Bernhard Pelzl, Wien, Köln, Weimar: Böhlau, 79–89.

Rinofner-Kreidl, Sonja 2013: "Husserls Fundierungsmodell als Grundlage einer intentionalen Wertungsanalyse," *Metodo: International Studies in Phenomenology and Philosophy*, vol. 1, no. 2, 59–82.

Rinofner-Kreidl, Sonja 2016: "Moral perception: On what it requires and how it benefits from reading fictional texts," *in Ethical Reflections on the Mind/Body Problem*, ed. Beatrice Centi, Firenze: Le Lettere, 307–348.

Rinofner-Kreidl, Sonia 2022: "Husserl's concept of the absolute ought: Implications for ethics and value theory," *The Existential Husserl*, eds. Marco Cavallaro and George Heffernan, Dordrecht: Springer, 251–275.

Rizo-Patrón, Rosemary 2019: "The ideal of humanity versus worldly powers: Kant, Fichte, and Husserl," *Revista de Filosofía Open Insight*, vol. X, no. 19, 61–80.

Salice, Alessandro 2018: "Practical intentionality: From Brentano to the phenomenology of the Munich and Göttingen circles," *Oxford Handbook of the History of Phenomenology*, eds. Dan Zahavi, Oxford: Oxford University Press, 604–622.

Scheler, Max [1913] 1973: "Ordo amoris," in *Selected Philosophical Essays*, ed. David R. Lachterman, Evanston, Illinois: Northwestern University Press, 98–135. Original "Ordo amoris," in *Schriften aus dem Nachlass I: Zur Ethik und Erkenntnislehre*, 2nd and revised edition, ed. Maria Scheler, Bern: Franke, [1916] 1957, 347–376.

Scheler, Max [1913/1916] 1973: *Formalism in Ethics and Non-formal Ethics of Values: A New Attempt Toward the Foundations of an Ethical Personalism*, trans. Manfred S. Frings and Roger

L. Funk Evanston, Illinois: Northwestern University Press. Original *Der Formalismus in der Ethik und die materiale Wertethik*, Bern: Franke, [1913/1916] 1960.
Scheler, Max [1926] 2021: *Cognition and Work*, trans. Zachary Davis, Evanston, Illinois: Northwestern University Press. Original "Erkenntnis und Arbeit: Eine Studie über Wert und Grenzen des pragmatischen Motivs in der Erkenntnis der Welt," in *Die Wissensformen und die Gesellschaft, Gesammelte Werke, Volume 8*, ed. Maria Scheler, Bern: Francke, 1960.
Scheler, Max [1931] 1976/1977: "The idea of peace and pacifism," trans. Manfred S. Frings, *Journal for the British Society for Phenomenology*, vol. 7, no. 3, 154–166. Original "Die Idee des Friedens und der Pazifismus," 1976 (Oct.) and continued 1977 (Jan.), posthumously published in 1931, Berlin: Der Neue Geist Verlag.
Sokolowski, Robert 2008: *Phenomenology of the Human Person*, Cambridge: Cambridge University Press.
Sorensen, Kelly and Diane Williams (eds.) 2018: *Kant and the Faculty of Feeling*, Cambridge: Cambridge University Press.
Sowa, Rochus and Thomas Vongehr 2013: "Einleitung," In Edmund Husserl, *Grenzprobleme der Phänomenologie: Analysen des Unbewusstseins und der Instinkte, Metaphysik, Späte Ethik, Texte aus dem Nachlass (1908–1937), Husserliana XLII*, eds. Rochus Sowa and Thomas Vongehr. Dordrecht: Springer, xix–cxv.
Spano, Nicola 2022a: "Husserl taxonomy of action," *Husserl Studies*, vol. 38, no. 3, 251–271.
Spano, Nicola 2022b: "Volitional causality vs natural causality: Reflections on their compatibility in Husserl's phenomenology of action," *Phenomenology and the Cognitive Sciences*, vol. 21, no. 3, 669–687.
Stein, Edith [1935/1936] 2002: *Finite and Eternal Being: An Attempt at the Ascent to the Meaning of Being*, trans. K.F. Reinhardt, Washington: ICS Publications. Original *Endliches und ewiges Sein. Versuch eines Aufstiegs zum Sinn des Seins; Anhang: Martin Heideggers Existenzphilosophie; Die Seelenburg*, published also as *Edith Stein Gesamtausgabe, Volume 11,* Freiburg, Basel and Vienna: Herder, 2006.
Steinbock, Anthony J. 1994: "The project of ethical renewal and critique: Edmund Husserl's early phenomenology of culture," *Southern Journal of Philosophy*, vol. 32, no. 4, 449–464.
Steinbock, Anthony J. 2007/2009: *Phenomenology and Mysticism: The Verticality of Religious Experience*, Bloomington, Indianapolis: Indiana University Press.
Steinbock, Anthony J. 2014: *Moral Emotions: Reclaiming the Evidence of the Heart*, Evanston, Illinois: Northwestern University Press.
Steinbock, Anthony J. 2018: *It's Not About the Gift: From Givenness To Loving*, London, New York: Rowman & Littlefield.
Steinbock, Anthony J. 2021: *Knowing by Heart: Loving as Participating and Critique*, Evanston, Illinois: Northwestern University Press.
Steinbock, Anthony J. (ed.) 2022: *Phenomenology and Perspectives on the Heart*, Dordrecht: Springer.
Trincia, Francesco Saverio 2007: "The ethical imperative in Edmund Husserl," *Husserl Studies*, vol. 23, no. 3, 169–186.
Trizio, Emiliano 2021: "Phenomenology, teleology and theology," in *The Husserlian Mind*, ed. Hanne Jacobs, London: Routledge, 521–531.
Vandevelde, Pol 2016: "Review of *Husserl's Ethics and Practical Intentionality* [by Susi Farrarello]," *Notre Dame Philosophical Review*, June 21, 2016.
Vanier, Veniero 2022: "Husserl and non-formal ethics," *Phenomenology and Mind*, vol. 23, 66–78.
Williams, Bernard 1981: *"Moral Luck": Philosophical Papers 1973–1980*, Cambridge: Cambridge University Press.
von Wright, Georg Henrik 1963: *Norm and Action*, London: Routledge & Kegan Paul.
Zahavi, Dan 2018: "Intersubjectivity, sociality, community: The contribution of the early phenomenologists," in *The Oxford Handbook of the History of Phenomenology*, ed. Dan Zahavi, Oxford: Oxford University Press.

Translator's Preface

There is an elegance to the writing style of Edmund Husserl. It comes not so much from an attempt to play with the sight and sounds of words but from an attempt to express systematically thinking itself. To that end, he would not only constantly work but rework, formulating previous thoughts, exploring new matters, and putting these in relation. Since most of the selections of this volume come from his working manuscripts, there is, at once, a fragmentary, unfinished, unpolished, and yet dynamic and masterful quality to them. As such, they present problems to translation, even beyond the usual questions of rendering complex German into readable yet accurate English. What is more, there came the responsibility of making systematic decisions about how to present the *ethical* thought of Husserl—something that had not yet been done in English scholarship, though there was important precedence to which to pay heed. So, in addition to the Editor's Introductions, a few general, then more specific remarks on the translation are in order.

As a general rule, the translation aims to offer the reader the text as if it were written in English, but in a way that leads the scholar back to the original German. For this reason, page numbers to the *Husserliana* volumes have been provided in-text, in their approximate location, so as not to disrupt the reader. A return to these volumes also provides the opportunity to refer to the work of the editors, whose *textkritische Apparate* are essential to understanding the history of the text. This text retains most of their additions and remarks, set in angle brackets.

More specifically, there are a few terms that, due to their significance and difficulty, demand some explanation.

Gesinnung is an especially difficult word to render in English. This fact is evidenced by the various ways it is translated in English phenomenological and philosophical literature, such as with "conviction," "disposition," "character," "mindset," or even with the phrase "basic moral tenor," and so on. Although there is an argument to be made for each of these, I have chosen the term "sentiment" for a few reasons. In the first place, the word derives from the Latin, sentiō + -mentum, where the former is related to the German *Sinn* ("sense") and the latter, like the German prefix *Ge-*, denotes either a collective noun or a verbal noun that may refer to the result of an action. In other words, *sentiment* is, in a way, a literal translation of

Gesinnung. In the second place, and more significantly, there is the fact that the term "sentiment" relates to "sense" and "feelings," and these relate in Husserl's analysis, as feelings have a role in how we make sense of ourselves and our world. So, in the third place, and perhaps most significantly, the single term "sentiment" captures the various senses of a specific feeling, a collection of feelings, as well as a general state or orientation, that is, one's "alignment": indeed, in these respects, it is possible to say that someone "has the sentiment" or even "shares the sentiment." Of course, there are drawbacks to the use of this term, not least of which its very ambiguity. In fact, traditionally, the term "sentimentalists" is used for what Husserl names the *Gefühlsmoralisten*, that is, those ethicists who, according to him, emphasize empirical feelings or emotions and overlook reason as well as the intelligibility and normativity of feelings and emotions. Husserl, neither rationalist nor a sentimentalist, means to restore the place of sentiment in ethics, however. So, I enlist the ambiguity of the word as one of its strengths.

Likewise, I render both *Gemüt* and *Herz* as "heart." As the "Editor's Introduction" to this volume makes clear, the latter tends to appear in a metaphorical context, such as when Husserl talks about values to which the heart can rise or a plan upon which one's heart is set, whereas the former has a more systematic use. Nevertheless, both denote a sphere of feelings or emotions and, as such, *Gemüt* has appeared, in other translations of Husserl, as "emotion," "emotive," and so on. In an attempt to retain the sense of a sphere of acts of feeling and valuing that relates to the logical acts of thinking and judging, I have chosen to use "heart."

Another important term for Husserl is *Besinnung*, which I render as "reflection." Like *Gesinnung*, it has its root in *Sinn*, "sense"; with the prefix *Be-*, it suggests a relation to or even a change in sense. So, through such reflection, one objectivates, say, operations of one's cognitional and emotional life, but one is changed through this becoming aware of oneself, and in such a way that one is aware how the sense of one's subjectivity exceeds any such objectivation. The term may be understood as a radical "sense-reflection," but using this as translation becomes cumbersome in the English, especially where Husserl uses a compound, like *Selbstbesinnung*, *Weltbesinnung*, or *Menschheitsbesinnung*, which have been, as a rule, hyphenated the English, as self-reflection, world-reflection, and humanity-reflection, respectively.

There is also the web of difficulties with "act" and "action" words. In general, we chose to render terms around *Handeln* with "to act," and so *Handlung* with "action." In the effort to avoid confusion, *wirklich* is rendered as "real" or "really," *Wirklichkeit* as "reality," *verwirklichen* with "to realize," and so *Verwirklichung* with "realization." In this volume, these decisions do not lead to any considerable difficulties with terms like *Real*, and so on. However, it does lead to the issue of showing how those connect with other German terms that take *wirk-* as a root, which I tend to render with terms around "effect," though I employ the term "influence" for *Einwirkung*. At first, the solution seemed to be to use terms derived from the Latin *operārī*, but this led to tortured English and conflicted with the use of "to operate" for terms near to *tätigen*.

For the term *das Gefallen*, I decided to use terms around the English "pleasing," though these are not to be confused with the terms surrounding "pleasure" (*Lust*). It is also worth mentioning here that Husserl uses the former in phrases where its sense is more like "doing a favor" for someone (*der Gefallen*). Furthermore, I follow precedence in Husserl scholarship and render the German term *Kunstlehre* as "technique." If only because the English word has suffered under the criticism of mechanical or rote technicization, it is worth emphasizing here that Husserl means it, not merely in a mechanical sense, but also in the sense that techniques emerge in the very operation of our living, which becomes clear in these texts. Likewise, the term *Gesetzgebung*, "legislation," should be heard in the sense of a "law-giving," through the operation of the subject in the world, and not a one-sided application of laws, either from within or from outside of the subject. Lastly, I have rendered various expressions of the *Ich-du* as "I-you." Although "I-Thou" is correct, the term "thou," which was originally an informal pronoun, and so carried the sense of familiarity, has since gained the sense of a *formal* pronoun, which is not what Husserl tends to mean by it.

With my thanks to the editors of this series and to all the others involved in it, let me also express gratitude to Mathew Clemons, whose insightful comments made this work not only better but also that much more enjoyable. It has been a privilege to work with them on these texts from Husserl.

Boston College
Chestnut Hill, MA, USA

Andrew D. Barrette

Contents

Chapter 1
Husserliana 42, Text Nr. 24: Value of life. Value of the world. Customary morality (virtue) and blessedness.

Edmund Husserl

<§1. The natural view of the world. The personal world and the world of things in their subjective orientations. The human being as a valuer and practical agent.>

<1)> The human being finds itself in relation to other human beings and finds itself among them as human, according to the lived body, soul, ego. The human being finds itself in relation to other human beings as "neighbors," namely, as members of a close-knit community. The human being itself belongs to the community and subjectively apprehends itself as its null-member. From the point of view of the other members, whom the human being apprehends as respective null-members by putting itself in their place, it is, in turn, a fellow-member. The human being is in relation to its nearest community and, at the same time, in relation to its neighbors as units of this community; and in relation to this nearest community, the human being operates as the central member of more encompassing communities that can themselves be the central members of even more encompassing ones.

The I-totality and each specific sociality in orientation. I and the others. *Each I-community "orients."* I—the other family members. We all (the family). We and the others; we human beings of our family—the people of other families. My family, our family—other families. Our family and the village community, the other families, those of the village community. Now *"our family" is the central member*

<Edmund Husserl, *Husserliana 42*: *Grenzprobleme der Phänomenologie. Analysen des Unbewusstseins und der Instinkte. Metaphysik. Späte Ethik (Texte aus dem Nachlass 1908 – 1937)*, eds. Rochus Sowa & Thomas Vongehr. (New York: Springer, 2014), 297–333.>

E. Husserl
Freiburg im Breisgau, Germany
e-mail: husserl@springer.com

S. Heinämaa, A. J. Steinbock (eds.), *Values of Love and Ethical Reflection*, Husserlian Legacies: Themes for the 21st Century,
https://doi.org/10.1007/978-3-031-68698-6_1

of the village community, both for me and for us who speak of our family; likewise, from the "standpoint" of every other family. In relation to the family, the village community is an objectivity of a higher order and, [298] like the family, an objectivity that presents itself in its "elements" as subjectively oriented in the schema "*central member and counter-member*." It then goes on like this.

Since every individual subject can find themselves as an individual person that is a central member of communities of various levels (I as family member, village member, citizen, and so on), every individual subject can also speak of its respective community; and, in doing so, the community itself has, in the individual subject, a consciousness of itself in a community-subjective orientation, and so on. *Thus, in this way, the human world as individual and communal is objective in its subjective orientation for individual human beings and for itself.*

2) A world of things is there, again in subjective orientations, for the human being and through the human being for the community. Things are experienced, but they also affect the heart and are handled. They are, above all, in-themselves and, even when they are experienced, valued, and handled by human beings, they have their in-themselves. At most, the things only "change"; and, if they step out of the visual and volitional sphere of the persons, they remain for themselves as something impersonal, as the "matters" that they always were. Matters are divided into those that are subjectively changed, actively formed, and those that are the way they are and have become the way they have without human intervention. This holds for each human being, for themselves: the human being's thingly environing world is divided into that which it co-forms and that which it does not co-form. And this is so for each community. Likewise for each human being and each community, with respect to the following possibility: the realm of things that can potentially be transformed by the human being and by the other things.

Lived bodies have the properties of matters; and this is so by nature. But they also have subjective properties, they are immediately under subjective influence, and so on. Human beings are integrated in the world of matters (the *res extensa*) through their lived bodies. Other subjects regard each lived body as a mere thing, nature, and, at the same time, as a lived body belonging to another; I can also regard my body as twofold and treat it accordingly. Lived bodies are the connecting links between both worlds, the *world of persons* and "nature" as *world of matters*. [299] Subjects have an effect on the world of matters in that they "animate" their lived bodies and, through lived bodily activity, which is at the same time a material happening, nature changes. Things, which among themselves are mere things, condition each other with regard to their changing and unchanging; they are dependent on the "causal" chain of things. The spiritual being-affected by things and <the> having-an-effect-on-things is a totally different matter, not to mention the purely spiritual effects of person on person.

If we understand everything as animated, then nothing essential changes. Then all matters (things) are at the same time lived bodies, just as lived bodies are also things in accordance with their essence. What remains is the difference between the effect-structures. Things, even if they are animated, have their nature and natural laws, their manners of effecting by push and pull, as the animal lived bodies do in relation to other lived bodies.

3) But the objectivity (of the real world) always includes something else. The human being has the world, the universe of real objectivity before itself as fulfilled time and correlatively it also has its universal (uniformly flowing) life, its being conscious, being affected, effecting. But the matter is not so simple: in the human being's conscious life, it consciously has the world as an open infinity, as a realm of the givenness of perception and memory; but every such experiential present <is> surrounded by an indefinite horizon. Each particular happening, each, has its particular horizon that is ordered within the total horizon. Every specific event, every specific action in this environing world is an acting into the future and into the future horizon. The future is the relatively determinate and yet indeterminate realm of expectation and of "real" possibilities. The prefiguration of what is expected and what is really possible is variously determined and rich, and it depends on the motivations that have been built up over the course of life up to that point. Also, even if the expectation is determined and subjectively certain, that which is expected need not after all occur. The subject is also aware of this, can become aware of this at any time. It can always be otherwise.

The related position-takings of the heart and practical position-takings: hope and fear. The future good of the expectation is that for which is hoped, the future bad is that which is feared. But the possibilities of turning out otherwise are not only phantasy possibilities. [300] Often enough, something other than what was determinately expected has come to be the case; and so what came out otherwise also has its "real possibility," in which there is a belief component of an analogous kind to that of the expectation.[1] So, to every hope belongs a motivated fear, with respect to the ready formed thought of something coming out otherwise, a fear that, instead of that for which is hoped, something to be feared will come about. But there is also that a better thing could and will come out—usually as a remote, only exceptionally motivated possibility—and so, too, a new hope.

To all of this belongs a general uncertainty of life. This is all the stronger the more the subject's experience has shown, in general, breakdowns of hopes and expectations. Thus, life is tossed around between fear and hope and filled with restlessness. In all action, the human being reckons with "chance," and, on the other hand, with that which is probable and probable in degrees, that is to say, with the empirical certainty that is, after all, only the anticipation of real counter-possibilities. To come to know the world, to derive advantage from experience and to proceed from the most comprehensive experiential knowledge of the world that is possible—this is the positive line of expectation and confirmation of expectation, the line of presumptive empirical certainty, to follow and to construe it. However, in contrast to this, there always remains the *infinite realm of vicissitudes*, of "it can also be and come out otherwise," and, accordingly, my action can miss its aim. This applies not

[1] <The editors of the *Husserliana* volume note that, at this point, Husserl remarks, "From here there is no clear train of thought." He corrects himself, however, saying: "It is quite serviceable after all.">

only to the world of things. It also applies to human beings, and in both a physical as well as in psychic respect, also to myself: my memory can fail at the crucial moment, I can become physically and mentally ill, and so on.[2] [301]

<§2. Valuation of life. The intertwining of my life with the lives of others.>

There arise other issues here. Every personal subject, every human being has not only the representation of the unity of its life, which is a constant experience, but also the experience of a certain typicality of life. Each has the experience of a normal and on the whole *satisfying life*, successful in its purposeful activities, despite inhibiting contingencies. The life is satisfying in the ascent to ends evaluated as higher and their successful attainment, in holding itself in a valuated height; it is unsatisfying when a *plus ultra* enters the horizon and when the level sinks or even when it eventually becomes monotonous (law of deadening).[3] "*Unhappy*" is a life that descends ever more, where nothing that can swing higher ever succeeds. Reasons for wretchedness: the unsatisfying life has its exterior and interior reasons, its reasons that lie in me and in nature and in the others, in particular individuals with whom I associate, in the context of the community, in the psychic, the social, and the physical, the lived bodily. Illness is a general title for lived bodily changes, and spiritual changes too, that cause unhappiness or that interrupt happy life.

Main titles: poverty, hunger, love, and death. "Struggle for existence" as competition of conflicting life-interests of different persons in the community of life. Spiritual love and *community of love*, in which several subjects live a life united in personal identification of the heart and the will.[4] What you wish, I wish; that for which you strive is also that for which I strive; what you will, I also will; in your suffering, I suffer, and you in mine; in your joy I have my joy, and so on. There is no strife here, but unanimity, not through concession, but rather through a habitual identification of the subjectivity of the heart and of the will in which a peculiar unity grows. [302] It is a unity of manifold life that is manifold on account of multiple subject poles which, however, are in a habitual coincidence in their personal acts, in the strictest sense, so that in each such act, multiple subjects have an effect and live it out in a describable manner: one as willing first, the other as co-willing, willing through the other will, and so on.[5]

---

[2] <Again, the editors of the German volume note that Husserl writes and then deletes in the margins: "Not of much worth.">

[3] Satisfying life. Unsatisfactory = showing a deficiency in satisfaction? Negative form: unhappy

[4] It would be necessary to insert the difference between reciprocal and unilateral identification of love, community of love.

[5] As a separate topic: not the community of love as a personal bond of persons with persons, but community of love of persons with family, people's community, persons and native city, and so on. Here the relationship is one-sided, just as unique-personal love can be one-sided (a one-sided

The unification of love multiplies happiness and unhappiness for each individual, but it also makes the unhappiness of the one only a partial unhappiness by virtue of participating in the happiness of the other. This happens when the life of the other still shows layers of its own happiness (or better <of> success, of satisfaction).

The peculiar occurrence of the death of the beloved and the unchangeable degradation of the happiness-value of the life that is impoverished and made lonely by the loss of the beloved. For every human being, the incomprehensible fact of its own future death, incomprehensible because it is not realizable from within, intuitively presentable—the uncanny great enigma, the uncanny great emptiness of nonbeing.

Valuation of life.[6] But my life is nothing in itself; it is united with the life of the others; it is a part in the unity of the communal life and reaches above and beyond it into the life of humanity. I cannot value my life without valuing the lives of others interwoven with mine. All living together is life that establishes relations of coincidence between an I and other Is with respect to life—relations of sympathetic unity or of antagonism, in aversion, antipathy, and so on. No life without love. And each life only becomes conscious in unity with a consciousness of love, a coinciding in love: the relation between parents and children. [303] Are not all values related to life and to the idea of a life that proceeds as valuable in relation to itself and valuable as a whole? Can anything be of value without relating it to a possible valuator and to what satisfies itself in the value-goal? And is this relation merely an empty relation of possibility? Would the world have actually gained a higher value if, somewhere on the moon, atoms assembled themselves by chance into a masterful "work of art" that no one has ever seen nor will ever see? Do undiscovered truths that will never be discovered have a value as cognitional formations possible in phantasy?

The full human being, awakened to humanity, values life in its universality, and, first of all, values its own life against the background and in the context of community life, of life engulfing itself in those nearest and then more distant.[7] And so the human being *necessarily strives for "blessedness,"* first of all, desiring it. It does not desire blessedness in any philosophically interpreted sense. We naturally cast aside all interpretations; we instead take the concept from the actual life itself and from its essential character. Life is in itself striving; as human, life is an ascent from passivity to activity, from letting go and being carried, from a life according to inclination, to one of deliberate choice, to a critical, higher-valued life of preferences.[8] That which is conscious of being inferior is no longer good, but rather is bad. The better is the enemy of the good. Necessarily, the human being arises to a valuation of life from the point of view of the best, the most preferable, but such that is a good and

bond), which has not been considered above. Non-egoistic interest in the other or here in the life, flourishing, welfare of the community. This can be extended by viewing the whole of humanity, finally, as a community of life and having a loving interest in it.

[6] The valuation of life takes place in unity with the valuation of the subject of life.

[7] The human being as human being does not live in the moment. For the human being, the whole unity of their past is a given unity that comes to be surveyed and grasped, and this applies also to their future and to their whole flowing life.

[8] Everything has not yet been thought through to the end. A bit too cursory.

can satisfy, that is best satisfying, that fully satisfies because no better is practically possible. Obviously, all this deliberating and choosing and valuing of that which is best relates to that positive line, that is, to the "hypothesis" that the probable will happen, that expectations will be fulfilled, and that, in a fixed and foreseeable environing world, [304] I now have the choice to seek out the best, and that I myself am stable in physical and mental health, in my ability, and so on.

I must decide, but at any moment I can err, I can fail in various respects, and, not only can I wrongly gauge the values themselves, but I can also wrongly gauge the course of the world, and so on. Despite all this, I act, I do not let myself be paralyzed, I put aside chance, fate, death, illness, possible failure of my own powers; I want to decide to the best of my knowledge, conscience, and ability; I want to "get myself together," I want to "collect myself," I want to do it in deliberation and then in execution. But how can I do that if I believe it is of no use? Have not the best succumbed to this thought and achieved nothing? Succumbed? They have not achieved their goals, the material ones. But if they have remained heroic in this will, when they have not allowed themselves to be bowed down—well, in that case, they have not succumbed and their life was "heroic" and worthy in the highest sense.

But how do the weak endure, when, weakened by ever-new fates, they cannot raise themselves to the height of heroism? And how do natural human beings, tormented between boom and bust, between momentary periods of satisfaction and periods of need, endure their life? By mere thoughtlessness. Need teaches to pray; and the greater the need, the more we look for a foothold. Here the mythical powers of tradition present themselves, formed already from the outset by the feeling-motives of need, fear, and so on, as forces of evil and good, the latter as salvific, benevolent, and so on.

<§3. Self-valuation of the subject. The form of the rational life. The dark horizon of senselessness.>

The valuation of life is, at the same time, the *self-valuation of the subject*: the human being makes itself the theme of valuation and the practical theme, including its whole life and immersion in the world, its suffering on account of the world, acting on the world; and likewise with others. [305] The human being evaluates life and itself as a subject. As a subject, it is free, provided that it resists passivity, provided that it can "overrule" practical affections, inclinations without further ado, that is, can deliberate over them and freely state the Yes and the No of its will. The subject evaluates itself in relation to its freedom, it reproaches itself, feels remorse or accepts its decision afterwards and is satisfied with itself. By considering itself a subject of a coherent unified life (in relation to the world that "stands over-against" its life) and by considering others in the same way, it apprehends the typicality of its individual personality, its personality as a willing personality, its character on the subsoil of its passive dispositions as well as its acquired habits, and so on.

The following question is accordingly familiar to the human being: How do I avoid self-blame and the blame of others, *how do I become good* and how do I, as a good person, shape a good, satisfying life, one that I can evaluate as good, with which I can be satisfied? I can will freely, I can will on the basis of deliberation, that is, deliberation over the circumstances of the particular case, the external circumstances, their probabilities and the possibilities, the consequences of each possibility, the values, the favorable and unfavorable consequences, and so on. But I can also estimate my powers, bring into account my capacities and skills as well as those of the others whom I must take into consideration. The others are factors in the external reckoning, but I myself as an agent have to posit myself as an interior factor. But it is not only in particular cases that I can proceed in this way. Alongside other general purposes and the considerations related to them, I can posit a *formal end* that encompasses them all. Surveying my whole life and evaluating it and myself as a free willing subject of this life, I can weigh *the general practical possibility of conducting my whole life rationally*. My life is rational and I am practically rational if I want the best possible overall and carry it out as best as possible, and if I set my will on not letting myself go at all, set it on striving for, doing, and wanting the best of all possibilities. Thus, I am not only, objectively and accidentally considered, rational, but consciously rational; I live in the sentiment of reason, and my life itself is worth more through this sentiment. [306] This deliberation is "ethical," it demonstrates for me the form of the rational life. However, it is not mere knowledge that makes me ethical, but rather free decision.

But how so, if life were worthless, *if it cannot overall become a life that is good through my efforts*, if it is *beyond my power to bring about*? How, if my free will does not reach that far or is even an illusion, if I were a plaything of blind forces that rule my life and the whole world, effecting in me the illusion of being freely capable. Must one not say, "Formal axiology and formal practice are, just like formal ontology, 'hypothetical'"? Namely, that something is and that I am and can do something, that I have goods in my practical sphere which I can realize, and can always gain something good and something best, that I can effectively institute the will to be good, to have within me a universal good sentiment, and so on—all this is not decided by formal ethics.[9]

Does not the "I can" stand at one with the "I am" at the beginning of philosophy? And even if I am very familiar with failure, familiar with occurrences that do not issue from the will or originate in it, am I not also familiar with that which comes from the will, with that which is effected by me? Do I not have the evidence of "I act," so long as the action proceeds within recognized limits, in an "immediate" area, as well as the memory of completed actions? Thus, only the open possibility of inhibitions, "distractions," disturbances, and that with each action, remains. *I have the evidence of being able*, the *evidence of the freedom at one with the evidence of the possible disturbance of freedom which nevertheless does not annul my freedom, and so on*. Can I myself not take these disturbances into account? Is the

[9] But does that not belong to the essence of the human as human? He is not, after all, a jellyfish.

vicissitude of the will fully unpredictable? Do I not, as a practical human being, constantly reckon with it insofar as I observe and learn from past experience and recognize that even chance has its regularities and can, at least to some extent, be foreseen and, in any case, can be taken into account according to its probabilities? [307]

I decide upon deliberation, not blindly according to arbitrary inclination. Occasionally, however, I decide despite the deliberation and against its result, according to inclination. Afterwards I am displeased and I consider how to avoid something like this; I want to avoid it altogether. Even this will does not unconditionally reign; and where I act against it, I have the consciousness of discordance, of a contradiction of will and of my own unworthiness and of remorse. I affirm my general will and my general valuation, I fortify its power, which can now overcome greater interior resistance. I progress, I reduce my personal unworthiness—in this way, I shape my ethical I, in this way I can shape it; it is a practical possibility. And when I experience that I can, then also I will, as a result of this "experience." I often do this, and if I do it, then I am free.

But does that release me from universal value considerations of the following sort? It may be that in the "I am" lies an "I have a practical field" and, therein, "I have a possible good." *But how can this be, if the achievable good continues to decrease*? How, if all the people I love pass away, if the masterful treasures of art and science are destroyed, if the humanity in which I live falls into desolate degeneracy, and the good I can effect in myself and in others becomes infinitesimally small compared to that which was already there, to that for which I could hope, to that for which must I wish? *And how, if I cannot gain any hope that it will ever be otherwise*? How, if I must assess the world as an *irrational* one, if the beauty of nature succumbs in natural upheavals and my estimation of humanity is lost, perhaps as a result of a war such as this one? *Can I live in a "senseless" world*? Perhaps I can recognize that the natural world has a system of rules that can be known by humanity, that humanity could create a beautiful world, if it consisted of reasonable people who made the world and their own life clear to themselves, the laws of the world and the sense of life clear to themselves and united. Yes, I can do this, if the lawfulness of the world has, down to the concrete, an order that does not allow the earth to become an icy wasteland in which everything freezes to death or does not make it into a Sahara desert, and so on, and if people agree to adopt reason and unite in accordance with reason, and so on. [308] But do I have any prospect of making people better, even if I myself am the reasonable one? Do I have any prospect of influencing them if they become mentally ill or something like that, if they act like idiots?

We have a history: higher cultures have grown out of lower ones, and there were living communities shaped by values at different value levels. But glorious cultures have collapsed and noble peoples have degenerated. And blind forces of nature have often appeared as destructive powers (earthquakes, the formation of deserts). Astronomy and geology also recount to us histories and show to us that the possibilities are not empty possibilities. Of course, from the ruins, a new life blossoms freely, at least often it does so. But a valuable one or one aspiring to higher values?

Is this world and this spiritual world a blind mixture of developments and destructions of value; and does each random mixture of present good and bad entail a small piece of rational achievement but such that it breaks off again and is interspersed with thousandfold irrationalities? Progress? Development?

Should we retreat to this: "Things are not quite so bleak now"? There is a culture with great values to which the heart can rise, and so can I and many other "educated" people. And influencing others is not so completely futile, in a similar manner as one learns from them. So let us try to conduct ourselves as reasonably as possible and to shape our communal life reasonably, as reasonably as possible, in this world, in this relatively regular persisting climate, in these lands that make regular local cultivation of the land possible and profitable, and so on. Here, reason does have a power. If I were, say, carried off as a prisoner to a foreign, lowly culture, unable to influence these people or bound in the lowest, what could I do? Of course, the "best among the achievable," and perhaps I could cast small lights into those souls that I find open. But I am in Europe, and a lot can happen here.

So, certainly, one can say: Here one can have an effect, find people with shared sentiment, exercising an influence that expands, a community of reason that grows. Does this not elevate the "true culture"? [309] And are not the people united here elevated by this? Here is a progress, a development that makes the participants not only objectively of higher worth, but elevates them in their own eyes and makes their life worth living for them, opens up to them an infinite open world that is itself open for the shaping and growing of values, values that are not sensually enjoyed but are enjoyed in elevating activity: science, art. And are further hopes then not possible? The outsiders will be able to see how something beautiful effects souls and how this beauty is ready for everyone, something that makes people happy and yet is a common good.

But the dark horizon remains. Is this all only for us and for this time and this probably transitory humanity? An accidental green spot in the wasteland of the world? But what do we actually know of the universe? Fragments. Cannot a universal science give us insights that expose such trepidatious possibilities as impossibilities? In the end, the wish is the father of thought.

<§4. Ethical action in the face of meaninglessness. Self-satisfaction and blessedness. Ethical communal life.>

What, in the end, is there to be said about all this? Let us assume that I believe—whether rightly or wrongly, it is all the same here—*that the world is "senseless,"* that human history does not and cannot bear in itself any progressive development, that the possibilities of lasting, value-laden effects and accomplishments for me and for all other people could not be reasonably probable. *What should I do then*? Should I do anything at all? The answer is clear. If I were a kind of achievement machine, contrived only to bring about something of "value," namely, value that lasts in a world of lasting values to be produced, and if I were the one who had set this goal for myself and could set no other goal for myself, [310] then the *knowledge of the impossibility of the goal would have to paralyze my will*. But how is it, when we think of the *mother who exercises loving care for her child*? She may know that the world has "no sense," that a flood that annihilates all "values" could come tomorrow; she may be convinced that, even if no conceivable probability speaks for it, there is no doubt that such a thing will eventually, "one day," come, and, if then a new world of values should arise, the flood will come again; and thus, no lasting value remains and perhaps, no certainly, the end is a chaos without any values. *The true mother will still say*: this may be so and even certainly so, but a thousand times more certain is that I must not abandon my child, that I should lovingly care for it and care for its lived bodily and spiritual becoming good and, thereby, its being good is an absolute value that I should set as a goal for myself. If I live according to this, then I myself am so far good and do my duty. That is good and remains good, whether the world still has a broader "sense" or not. –And if the mother "knew for sure" that tomorrow or in one hour the end of the world would come, would she fail during this hour, as a true mother, to give her child loving care, consolation, and so on? *And is it not similar with everything ethical*? People on a ship doomed to sink, knowing this, will still be able to act ethically as genuine people, in the work of love between each other. And all the more for genuine people who have before them greater horizons of free effects, like if earth were, in their opinion, a necessarily sinking ship destined to perish. Not because I can believe that the values that I create are eternally preserved or serve as the means for enabling higher values *in infinitum*, so that the world would be a museum of values my actions create and the actions of my fellow people or an infinite sequence of ever higher values—it is not for these reasons that my respective value is valuable. My value acquires the character of a *means* only if I have this belief and set myself the corresponding goal of collaborating in the infinite development, but even then, it is not merely a means, but value in itself, if it is value. The world may be a hell, it may be that the valuation of the universe brings forth no final sum of positive values, no lasting positive surplus of values, no lasting values; nonetheless, [311] I resist this hell and do my "duty."[10] I am and I can want something good, strive practically, even if only in a narrow sphere of influence; and if I do something good, then I have done my best and cannot reproach myself in any way. In a certain way I am satisfied—with myself.

But I cannot consider myself "blessed." *Thus, the categorical imperative comes into tension, that is, self-satisfaction and blessedness come into tension—blessedness when correctly understood (or "virtue" and* "blessedness"). The center of blessedness is self-satisfaction, the correlate of the best possible personal life that comes from the sentiment to will the best possible and to do it to the best of one's ability. Without self-satisfaction, then, no blessedness. On the other hand, blessedness is more than self-satisfaction. It is satisfaction with my natural gifts, my health,

[10] But can the universe in which a good lives, a noble mother, and so on, ever be a negative value? And is not every good deed itself in a certain sense an eternally "enduring" value?

and so on, and satisfaction with the environing world and so with the world in general as my field of effects.[11] The world can offer me higher or lower possibilities of happiness. The best possible that I create in it is of varying levels of possible value, depending on <how> the world itself is. In all circumstances, the world is pre-given and one among possible worlds. Each of these possibilities is able to be of value to me. Thought of as a world of this kind, it is at the same time, if as evaluating and acting in it, that is, looking at the world as a personal environment, it is a potency for possible free transformations and harbors in itself a potency for a best that could come about through me, and so precisely that which would give me self-satisfaction in relation to it.

But the situation is essentially more complicated. If the environing world were a world of mere things that I, just as I am, subject of these factical capacities to shape into a best possible world of value, or rather, world of goods, then it would be the end of me. But my environing world is "also" a *personal environing world*, in which there are persons like me, persons whom I can indeed treat as things but who nevertheless remain persons. As persons, they are subject poles of an evaluation originally related to them and not to me. [312] If I am not to violate my self-satisfaction, but to gain it, then true values for them and, ultimately, that which is the best possible for them and for their self-satisfaction is not a matter of indifference for me. Everything that is of true value for them is taken into account in my value calculation as a positive factor, inasmuch as I have influence on it, can support or disrupt it, or inasmuch as it supports or disrupts me in my own values that are originally related to me. Where the values of others lie outside of my practical sphere, they are, insofar as the others realize them, objects of pure joy for me. But the same is true for the others in relation to me. We do not, however, stand merely in relation to each other with respect to the material values that lie between us that are or can be shaped by us as material goods; and the disruptions do not refer merely to changes in the world of things by which we disrupt the possibilities of material production of others or they disrupt ours. It is much more that we are in communication through social acts. We live not only nearby each other, but in each other. We determine each other personally, from person to person, from I to I, and our will does not reach out to the other as things of the environing world, but reaches into the others, it extends into the other's will that is at the same time the will of the other and our will so that the other's deed can become our own deed, even if in a slightly modified manner. We work with one another, and that always involves acting through one another and in one another. We unite in common deeds as we did before in common plans, resolutions, actions, and we produce communal works. We have an effect on each other, we do it in praise and blame, in the effective consciousness of the overarching concordance or conflict of the directions of our striving and willing, in love and hate. We might unite and can unite in a common ethical sentiment, in the unity of ethical

[11] The world is the totality of my organs of action, my means of action and all real preconditions of my action and all action.

orientation of will. We do not each live our own solipsistic ethical life, but a common, doubly-personal and yet unified ethical life.

We have *a communal life* first through current empathy, and it is all humans that have the potential to empathize inasmuch as that they "communicate" and inasmuch as the same material world <is> there for all as the environing world, the selfsame nature, in which everyone can intervene and through which everyone in every intercession co-determines the passive life of the other in terms of what the other really experiences and can experience. [313] And each of their interior lives and each productive creation belongs to the life of the other insofar as each one is in the "world" with their interior life and is accessible to everyone through bodily expression, accessible also in their interiority, though in varying degrees of factual accessibility. On the basis of this common subsoil, through which each life reaches into the passivity of every other life, creates pre-givennesses for each of the others, the personal life of community takes place. This is the life in which, not only do particular individuals "communicate" with isolated poles of action as on the lower level, but in which, at various levels, and possibly only in occasional stretches, pole systems and correlative unities of life are also constituted with the character of a life that entails no isolated poles, no one acting for themselves independently and unconcerned about the others, but several poles that "participate" in each action for which all are "responsible." A special case is that of the *synthesis of poles*, that is, *personal synthetic unity out of freedom*, in which each I freely connects with the others and, as ethically free, perhaps enters into a "*community of love*" with them, freely making the values of the others its own or acknowledging and appropriating them in spirit and in truth as its own, and the others, for their part, do the same. As correlate, an ethical community life arises and in it a communal realm of goods that is the best possible for this community. As a community, this personal, synthetically linked multiplicity then gains self-satisfaction in which the self-satisfaction of the particular persons is then contained.[12] But this is an ideal case, the possibility of which must first be thought through. Prior to it are the cases in which persons live their specific lives, though they communicate with each other throughout or for the duration, they come into personal contact, and possibly in ethical relationship, only for certain purposes. [314]

<§5. Ethical community as personality of a higher order. Value possibilities in the face of human finitude and irrational chance.>

As a human being, I have an environing world. At the same time, I know myself and experience myself as a member of this world; I belong to it, complementing it; I experience myself as its "central member," as the null-member of an experienced

[12] Ethical community, ethical communal life and action. The virtue of community that judges itself as community and is self-satisfied.

world oriented around me. Yet, through empathy, I experience that every other, given to me in the orientation form of the *alter*, is given to themselves in the form *ego* in which everyone else has the "appearance form" of orientation of *alter*. If I build the idea of the world itself, as it itself is, so that everyone must recognize it as the same experientially and rationally irrespective of the different manners of orientation and other manners of givenness that it has, for one in that manner and for another in a different manner (but in a manner that is formally the same for each, according to the formal description above)—then I shall say: the world itself is one and the same for everyone. And it is nature for everyone, nature that contains our bodies as things that appear differently to us subjects in the alteration of subjective orientation and contains the I-subjects with their lives, the subjects who express themselves in lived bodies or are given in lived bodies in determinate order and regulation. Each is in itself what it is for itself in the mode *ego*. We thus have a nature and in it an I-wholeness in determinate arrangement, a totality of egos, namely every single I, even while the form of the *alter* belongs to the form of appearance of the alien body, and the I for me belongs to the appearance form of my own lived body.

From the absolute transcendental examination, the I-totality is absolute such that nature is only a unity of appearances and a synthetically uniformly regulated system of appearances in every I. This constitutes the existing nature for this I and also the experience of other egos in the form of appearance of *alteri* via the forms of appearance of living bodiliness and the experiential motivation of expression in lived embodiment. Such *alteri* are thereby experienced as modes of givenness of other egos, related to the same nature as I experience, nature as the identically existing unity of sensory experience. [315] Then the I-totality becomes absolute being, which is being for itself, and nature is the identical pole-system of appearances that we bring to synthetic intersubjective coincidence through empathy. As a human being, as a personal subject of an active life, reacting against passive affections, as an ethical subject submitting my life to a norm of the good, <I> have in my horizon this universe as the field of my effecting. I submit my life to a norm, that is, I submit to a norm my living within the world, my effecting within the world. *But in this practical horizon, <I> find others* whom I can have an effect, who have an effect on me and want to have an effect on me, with whom I can have an effect together, as they perhaps are willing to do or are directed by me to do. And I can have an effect with them both on nature communally in general and <also> have an effect on third parties that belong to the environing world for us communally. Possibilities arise there of a communal life in the sense of the levels of social unions, of the links of I with I, of several Is to pole-systems of real and possible activity, that is, originating from the individual subjects in their freedom and in social acts <as> activities directed at and through other egos. There arises the possibility of an ethical synthesis in which each I lives in an ethically effecting manner, and thereby realizes its best possible life by realizing the best possible for the others at the same time, but in such a way that it lives not apart from the others but in them (wanting through their wants, valuing through their values), and—in understanding and linking together with the will of the others—creates a community of action in which the connected egos as connected subjectivities become the "synthetic pole" of the actions of community. We have then an *ethical community that is the analogue of an ethical personality and a personality of a higher order*, and have the individual persons within it. We have a communal ethical life of a synthetic personality as an ethical personality, and we have an individual ethical life of individual subjects, both being, however, in one another and one with the other. It belongs to the categorical imperative of the individual subject to strive for this higher form of community and this higher form of individual being and to strive to become the *individual life as a functionary of an ethical community*. [316] Insofar as it is practically possible, it is itself of unconditionally higher value than individuals' living beside and beyond each other, and thus is categorically demanded.

All *higher values* lie in the self-activity of subjects as acting subjects. And we have correlative values corresponding to the correlation between ego, act <and> the formation generated in the act. A mere thing has no value. It can have value only in relation to what it "means" to subjects in terms of reality and possibility, what it means in possible valuation. As work it has meaning only in that it realizes values and in that I and others can recognize the meant value and understand and value according to it. The recognition of a value, however, means that I legitimately value the matter at hand—as giving certain manners of appearance under familiar circumstances. Thus, I originally value the manner of appearance and value it as idea (independent of my momentary lived experience) or the matter as a work *with* this purpose and this sense, this generated manner of appearance in relation to persons; and yet I do this possibly indifferent to the fact that it is precisely this person who created it, and so, again, that I value the manner of appearances as a certain idea, and so on.

The *highest values* are everywhere those of subjectivity as such, which is directed to value creation and value commitment and directed above all to the best possible. Blessedness as the joy of doing the best possible in relation to an environing world that allows doing the best possible in the second sense.[13] *The subject of self-satisfaction does the best possible in relation to the given environing world.* The best possible environing world is the one that allows persons to do so much good in the best possible deeds that no other possible environing world allows them to do anything better. Belonging to the value of the environing world is also the *most perfect world*. A world that sets the greatest tasks for humankind; and it can only set humankind tasks that are fulfillable. Such a world offers the greatest possibilities for the realization of the greatest values. [317] It enables the greatest possible socialization and, in this form, the possibility of the realization of values of the highest form, values of sociality, values of the community of love.

Values of development. There can be no death of values, no dead stabilization of a level of values with always the same style of values. There the law of deadening

[13] The passively presented amenity and beauty, which is itself material for transformation; and the better the substrate, the better the higher practical possibilities to be realized.

would come into play. The value level that remains rigid is *eo ipso* a sinking value level.

The powers of human beings are finite and the individual performance of a human being is limited. Possibilities of increasing the effective power of the individual subject and thus the level of beautiful achievements and blessedness. Medicine. Means of prolonging life (maintenance and possibly pushing back the limit of aging decline). With respect to the social value possibilities, humanity's constant population increase is of importance. The more people, the richer the social possibilities, the greater the tasks, and the greater the values. The increase in spiritual strength. Improvement of health as a physical fund of strength for spiritual achievement. Pedagogy, the higher rearing of spirituality through higher development.

The inevitability of death accepted. The endless chain of generations and the continuation of one's own life in descendants, linking together through originally instinctive bonds of love. The thought of death loses its sting, death loses its disvalue, if I know myself to be part of an endlessly continuing humanity that can freely interact with itself and build itself up into an ethical sociality.

Illness, chance, the irrational, the being torn away from great tasks. The irrational intrusions of natural forces into the rationality of the life of humanity. The struggle with folly, with madness, with egoism. This too, in many cases, comes under the concept of chance. A great and beautiful plan, a great life's work rendered void because the rational life of the good man collides with the egoism and the wickedness of individuals. The war to which, above all, the best fall victim.[14] [318]

Theodicy, cosmodicy. Eidetic consideration of ideal possibilities. The ideal of perfect knowledge and mastery of nature: infinite task and infinite progress.[15] *The ideal of more perfect human knowledge and human "mastery."* That is to say, human action, the capacity to raise people and to educate them in such a way that they allow themselves to be governed by motives of reason and, on the other hand, such that they can use their reason to an ever-increasing degree, can gain ever-increasing rational knowledge and rational achievements, and can posit as an ideal goal and gradually fulfill *the ideal of ethical self-rule and ethical life*. Ideal of progressive ethical socialization.

---

[14] Faith in reason. If I have even the least real possibility that the world is accommodating to human purposes, then I must take this presumption as a certainty and act accordingly. That's how I, at any rate, do the best possible. The idea of God, of a divine world. The ethical "as if." The faith that has its power from the ethical will.

[15] Perfection on all sides through human freedom, in infinite progress. The most perfect world as idea. Can I believe in it? Can I prove the belief? Can humanity, and to what extent can humanity realize it?

<§6. The idea of the human being as rational. Ethical individual type, individual categorical imperative and formal categorical imperative.>

The examination of the essence of the rational being (of humans) as finite being in relation to the world experienced by them, experienced to a limited degree and surrounded by open indeterminate infinities. The individual human and humans in their environment, the world that is immediately experienced and really accessible and partly known mediatively on the basis experience.

Examination of the essence of the human with respect to human achievements. Consideration of practical reason. The human as a practically rational judge of itself and its possible achievements, its values, of good and best purposes who then determines itself to be practically rational.

With respect to the essence of the possible human being, the finiteness of abilities and the development of these abilities to better capability. Finiteness of this development and yet without fixed limits. Also changing, as with the particular abilities themselves, from human being to human being. [319]

The communal world of human beings and the coexistence of their developments and their possible rational developments in the sense of individual practical reason. Each human being is dependent on their neighbor, dependent on the possibilities of their own best capabilities, on their own rational-practical sphere of their achievements, that is, dependent on what they must categorically, according to the determinate content, demand of themselves. Human beings are also all dependent on the course of nature. Nature as running according to its own laws, "by itself," even where humans do not intervene in it. Humans can and do constantly intervene in it. But its own being and its manifold events taking place before the intervention and its physical causalities are presupposed.

The human being adapts its praxis to the experiential knowledge of the world. Human beings extend this knowledge themselves and learn from others' experiences and experiential knowledge. Thus, the human being's practical horizon constantly changes by virtue of expanding and always constantly changing knowledge. Also, if we take this knowledge as ascending, then the human being's practical supreme goal becomes, in content, an ever higher, richer, more valuable one that is at one with that of other humans. To a certain extent, ascension is something normal. We all form a normal idea of humans as a type of similar kind as is the type of the organic normal idea of a tree and its whole development. This in contrast to the normal idea of the rational human as ideal and as ethical demand. The human grasps the type "normal human" and, as ethical, wants to enhance this type. That, in communal striving. However, human beings behold their individual type for themselves, that is, themselves as the subject of their given dispositions, character traits, with their own typical way of acting, of being motivated, and the type of their acting life. Such a human being, as an ethical human being, wants to change and to transform this individual type in the sense proper to the ethical norm, to give their individual type ethical shape. *In ethical humanity, every human being would have the general ethical type in common with everyone, but each would also have their own individual ethical type.*

If the human being is surrounded by infinities that, all taken together, are called "the infinite world," it is to be noted that they are infinities into which he can knowingly advance. [320] The necessary orientation in which the environing world is given signifies a general formal structure of the manners of givenness, as well as the structure of the permanently available ways for *knowledge* and for control motivated by that knowledge or for suitable concern of control. The world gives itself as existing world in constantly presumed certainty, in the progress of knowledge by virtue of a concordant, unfolding experience which advances in form of enriching cognition. This holds for nature, considered *in abstracto* and considered as mere physical nature; it holds for the corporeal and the psychic being of humans, for their psycho-physical "nature," and holds of each individual human being in the context of physical nature. Here, on both sides, prevails an *a priori* necessary ont<ological> form to which all possible facticity of real physical and animal and human nature is bound. Considered absolutely: The I-subjects as such—the monads as monads—have their necessary essential form, their general structures that can be shown <as being> necessary in every facticity: their necessary shapes of possible development, their necessary forms of possible free activity, their forms of possible factical abilities—passive and active—and their necessities concerning their coexistence, their common nature and the form of corporeality. This includes the ontological forms of every nature and the ontological forms of every corporeality and of all ont<ic> relations of corporeality, nature, and ont<ic> subjectivity.

Therefore, to the idea of the human and humanity in relation to the environment and the world belongs the *a priori* of the world. Also belonging to this idea is the possibility of a developmental form "human being," in which the human being is one who knows nature and in general practices world-science, in which the human being can know the world as existing and can know the conditions of the possibility of the being of a world in the system of ontology and, through it, can explanatorily know the given world in empirical-rational science. The striving for such science and the happy striving in individual disciplines is <a> fact. Yet, for the incomparably greater part, the given humankind does not know anything about the world or does not understand it, having only inauthentic and wrong representations of it. [321] And yet <it> is also <a> fact that every human of any normal level of elevation wants to be happy and act rationally and that striving for ethical reason is widespread. Likewise, it is a fact that *in concreto* it is recognized that there is good and bad action, appropriate and inappropriate, but also absolutely good, absolutely required in the given situation, and accordingly also absolutely reprehensible, bad.

We can see that, *in concreto*, every human being *has their duty in the given situation, their individual-concrete categorical imperative,* related to their practical sphere as meant and, on the other hand, also related to their real practical sphere as real in the sense that sphere which they could have discovered with their respective ability, their respective previous formation, in the case of the best exertion of their powers.

To the normal human being also belongs, as normal, the experiential knowledge of the endlessness of their environment as an incalculable one and as an open *realm of unreckonable vicissitudes* which interfere with their carefully considered calculations that are bound by the practically limited knowledge. This does not change the current individual categorical imperative or the formal categorical imperative for an ethical life, namely, to bow universally under the categorical demand or rather to recognize it joyfully—the demand to want to do the best possible in general and to do so in every moment of life, in a freely choosing way, as well as to give the unified style of rational life to their whole life, if possible.

<§7. Anti-rational fate and the struggle against fate. The imperfection of the world. Science and praxis.>

We now have, however, the essential possibility of a human being and a humanity that knows themselves, not only in a world that makes rational action possible and, for the individual, a rational-moral order of life in general possible, but also <in a world that,> as a realm of irrational, unreckonable vicissitudes, disturbs such practical reason. As such, the possibility that the unforeseeable vicissitudes accumulate and deprive or threaten to deprive the human being of the hope of being able to be rational. [322] Admittedly, this is not to be taken very rigorously. In every moment, the categorical demand remains with the human being, and they satisfy it when they strive for and initiate the best possible; even if it is unsuccessful, *they are legitimated before themselves*. But if they again and again miss the joy of success, they *cannot be satisfied*. The human being's striving has always brought about a good and at very least the high good of agreement with themselves, of self-satisfaction, but always less good has been brought about than was foreseen; and if it had become the better, it would have become so within the ethical form and within self-satisfaction. The artist cannot be satisfied when, despite his best willing and striving, he sees his work fail on account of interior or exterior irrationalities. On the one hand, the *dissatisfaction* eventually becomes dissatisfaction with oneself, not with oneself as the subject of the ethical sentiment and the operative ethical will, that is, as a specific ethical I-pole, but with oneself as the subject of deficient empirical abilities, as a human being of deficient nature (aptitude, ability, level of empirical skill). On the other hand, the dissatisfaction with *the factual world, which as hostile fate,* brings the best plans to nothing.

How to overcome such dissatisfaction? Should I take comfort in the fact that, if chance hinders my success, it can benefit others? Should I take comfort in the fact that my failure is not a general failure for everyone, since I see that others are fortunate where I am unfortunate, that the world, which is now and almost as a rule hostile to me, is friendly to others? But even if I am full of love for humanity, I will not be satisfied with this, if I do not see a *necessity that, all in all, everything is arranged for human satisfaction and that everyone must occasionally do without because only in this way can others be satisfied*. Even then, is this comfort enough? That is to be considered. But even now, one could say: if I see from general experience that everyone must suffer "fate" but that everyone *on average* nevertheless achieves their goals again if they remain upright and strive on, then I can compose myself and let storm and tempest pass over me, wait and strive bravely on. [323] And so I do not just do my best, what I possibly can do, I also thereby increase my strength and at the same time become a worthy exemplar for the others, like I become an exemplar for myself in a certain way: namely, in future dangers, I draw new sources of strength of courage by thinking back from past bravery.

Now, of course, I do not know whether the world will not one day prove so hostile to all humanity that it will annihilate all its striving, letting, at least on the whole, the sum total of happiness disappear in comparison with that of unhappiness. I do not know whether this will happen such that the estimate then showed that, on average, valuable striving for higher things fails and people, if they remain alive at all, remain bound to low spheres of value. But so long as no probability speaks for that, as long as the empirical generally shows instead that one can successfully live for greatness and beauty, then I must not despair. Yes, I will do the best, I will practically overvalue the probabilities and act *as if I had the certainty that fate is not in principle hostile to human endeavors*; I will act as if I could be certain that, by perseverance, I can eventually achieve a good such that afterwards I could be very satisfied with my perseverance. What is *theoretically reprehensible*—the overvaluation of probabilities or even just slight presumptions in favor of one empirical certainty—is *practically good* and thereby *demanded* but only in practical situations. Look to that which makes you strong! Believe in the world and in fate! Take the fate as if it were certainly a good one, and live as if you can eventually make it serve you, as if you can influence it, make it friendly to you! Live in this certainty and you will do the best! But that is to say: Bracket the unavoidable dissatisfaction when frequent dangers and disappointments occur, leave out the empirical expectation that the next categorically demanded striving will turn out the same!

But in the sad present, does not general experience speak for a hostile fate that destroys human happiness again and again and makes fruitful work impossible? But if the human being perseveres even there, fights heroically and strives for the good, it is nevertheless best. Only in this way is the human being *ethical*. But are they *satisfied*? [324] Why must they be satisfied? The human being must be capable of being satisfied in order to be able to set themselves the goal not only of being good, but of becoming better and better. To be good is to follow the categorical imperative. *But the human being grows with their higher ends* and the *progressus* of higher ends must be open to them. Human beings must have possibilities, in this respect, and it is precisely by directing their will to them that they have probabilities for them.

What are the highest *ends*? Realized values, which are not only lasting values for me such that I can enjoy them and elevate myself again and again by them, but rather those values that are values *for all others and for all later generations*, at least insofar as possibilities perpetually arise for others to have the conditions of development, in order to be able to value according to these values. Because of my love of humanity, which I must have as an ethical human being, every value gains for me a general human value, that is, a value for every rational being, such that I can understand according to it. This also increases the value itself and, at the same time, increases my joy in the thought of all the joys that the value is called to bring about. *The ideal would be that I create values that could function as elevating and as source of happiness perpetually into infinity*, thus having the "immortality" of the values, not only of their ideal possibility of being able to have an effect, but in the endless reality of the effect of the values *in infinitum*. But here belongs the fact that values are perhaps precursors for the obtaining of higher values that absorb them in themselves, but in truth found them. Thus, the value would have immortality in the form of being eternally efficacious, without having to preserve itself. If we add to this the value of the loving union of the people and the communal work on values that are eternal and always effective, then the *ideal of an ethical humanity, living forth and cultivating itself into infinity* is such that, if we think of its possible realization, it would be well-suited to give joy and, with joy, also momentum to human striving. A world that opened up this prospect would be a beautiful world despite the vicissitudes of fate, even if it would not necessarily be the most beautiful world. Not the most beautiful, so long as the vicissitudes in it would have to be regarded as irrationalities that stood as inhibiting, as value-degrading moments in relation to the progress in the creation of value and [325] the increase in value of the world and of the people themselves—and these as subjects of the creation of value through which they gain personal elevation.

Now several things come into consideration. First of all: We do not live on a Robinson Crusoe island, and the earthly environing world of humans is already for us a kind of endlessness in accordance with coexistence and succession. The expanses, including historical ones, in which we have an effect, are so large that it is, practically speaking, already as if we have an effect in an endlessness. Practical possibilities, despite the contingencies that determine whether they are fruitful, are generally there and probable horizons of the distant effect. The distances are unlimited, unlimited above all on account of not knowing whether they are finite or endless.

Furthermore, the great, value-raising meaning of the struggle against irrational adversities as heroism in perseverance, despite accumulated misfortunes. It increases value in the case of the individual, but also for the community because of the example it sets for others and the increase in strength they experience on account of it. There is, however, a great difference between the struggle against "fate," against the adverse vicissitudes that prevent the rational and ethical *sentiment* from having an effect, and the struggle against temptations, against sin, against logical, axiological and practical temptations. The motivations at issue here, the possibly necessary foundations, would have to be subjected to a separate analysis of essence. Is heroic virtue thinkable without temptation, without decline into sinfulness, without working out defeats and out of a heroic will to live that possibly follows? Is wisdom possible without error, holiness without moral sin? The field of freedom and bondage and what they presuppose in terms of their essences must therefore be dealt with in their own right. Irrationality has a wholly different sense here.

Furthermore, one can ask: to what extent can the *progress of the knowledge of nature* serve to constrain the realm of "fate" understood as the realm of the world's senselessness, its senselessness in relation to the authority of human reason? Constrain? Certainly. Control over nature, in any measure, is possible foresight. But what are the essential possibilities here, and what the essential necessities that delimit those possibilities? What are here the *a priori* possible ideals, what the practical, ideal end-ideas? [326] Here the *problem of the finitude and infinity of the world* comes into view. Is an infinite world at all thinkable? In other words, is the world necessarily finite? But if it is finite, then a finitely completed science of the world is thinkable? Here nature and spirit are to be distinguished. If the subjects were passive, if it would be thinkable that they all were passive, then we would have mere nature, and as finite nature it would be completely calculable. But one must not forget the subject who scientifically researches into nature, who belongs to the world and changes it through experience and thought, and also physically changes it by virtue of experiences in lived bodily activities. So, is not the finitude of the spatial world at the same time an endlessness of the occurrences in time and an endlessness that is incalculable? But must not the world be determined in itself and must not, in accordance with that idea, the science of the world in itself be determined?

Here, a manifold of essential problems are to be posed and treated. And this indicates the following: beyond the natural science of the world—natural sciences, human sciences, zoology, and so on—philosophy must be pursued and, ultimately, a universal philosophy must be built up such that it encompasses all possible sciences in a unifying manner. Such philosophy stands before our eyes in advance as an endlessness. But we need not yet know anything of its real endlessness. Here, as everywhere, we must distinguish between, on the one hand, *the endlessness of the horizons*, the consciousness of their indeterminacy, the possibility of opening them and the ability to penetrate into them in ever-new steps which lead to ever-more indeterminacy, and, on the other, the real endlessness. For the question presents itself as to whether we must essentially come to ever-new openings with new determinations—we, ideally spoken. This is also the case for the endlessness of the spatial things in their spatial distribution, <and for> the endlessness of the causalities and properties of the individual things. Even if we know that ever-new things could be discovered and, therefore, that ever-new causalities would have to be considered, it could still be that there are factually no further things and thus no further things to be factually discovered and so the world would be finite. So, a cognition would be thinkable, at any rate, that would no longer find any positive experiential motives for asking questions beyond a realm of things finitely closed and already brought to light. [327] However, we are now set up in such a way that our knowledge of nature has led again and again to new celestial systems and, within the already given spheres of things, <to> ever-new elements and intermediately real thingly materialities, and so we must empirically expect that it will go on like this. It would be something unheard of in astronomy if, upon constructing instruments for observation that extraordinarily enhanced our penetration into space, we would be forced to state that, behind all the stars that have already been discovered, the expanded view shows no more new ones. *De facto*, therefore, we have an openly indeterminate endlessness before us, similar to the way that we as pre-scientific people can endlessly proceed from the firm ground, on which we stand, without being able to presuppose a final step in expectation.

Our practical horizon does not look so essentially different—with regard to its form—from the practical horizon of pre-scientific people. Pre-scientific people, too, have their experiential knowledge that looms before them in these indefinite expanses, and they "calculate" the practical possibilities among which they must choose reasonably. In this, they also have their neighboring people and, for that purpose, they are served by experiential knowledge—just as they are with regard to physical things. In principle, of course, they had no more psychology than this available.

But everywhere the following is valid: *the better the knowledge of the world, the better possibilities for acting*, however vicissitudes break into it. Even if cognition were free from unreasonable inferences and cognizing happens in the relatively best possible form, all calculations can be dashed by an unforeseeable fact. *Science makes foreseeable that which was not foreseeable before.* It cannot be otherwise. On the other hand, it *enables better praxis*; it opens up greater and more rewarding tasks. *It becomes itself an open realm of the praxis of reason* with exactly those vicissitudes that destroy previously designed scientific plans, overthrow theories, and so on, and yet make progress possible. And this praxis of reason is, at the same time, *a means for each other praxis of reason* and can at least become so. This *gives possibilities for a relative satisfaction.* [328] I belong to a world in which I can set rational goals for myself and in community with others and can work joyfully. In community with others, I have an endless horizon of fruitful work, or at least an empirically endless one. I do not know how the world will ultimately be, I do not know whether it will always be so and must be so. I do not know and we do not know whether a world catastrophe will not suddenly put an end to all striving. I know empirically that I will die, that my personal labor and achievement will come to an end, my personal happiness, if my success bestows happiness upon me, is a temporary fact. But I am easily consoled by that thought that, if I love others, *my work is a link in a chain of work* that continues through the chain of generations in the framework of the endless world reality, and that what is good in it benefits others and—their having improved, heightened, widened it—again benefits the coming generations. The horizon is such a wide and open one that I still have a certain *relative satisfaction*, even in the face of the possibility that this horizon is perhaps only a finite one. But I know nothing about that. That is something that does not allow my satisfaction to be perfect. If I believe in advance in the finiteness of the continuity of generations, this does not cancel my ethical striving, but I must then judge the world to be *imperfect.* Not as worthless, for it bears values and values that are still increasing, but it has an infinite lack when compared to the necessary ideal of an increase *in infinitum.*

On the other hand: I can collapse or break down; it is possible that I become factually inhibited in the best expectation, that the most beautiful and noblest plan fails, one upon which my heart is set. And how many true misfortunes I see around, how many disappointed noble hopes! That is a real lack here, at this place of the world, and is my misfortune, real misfortune insofar as it concerns me. I cannot be satisfied with this if I let my gaze rest exclusively on this. Still, I cannot reject the world, so long as I have to acknowledge that the vicissitudes are so distributed, according to general experience, that they do not cancel out the possibility and probability of the world's general value achievement and value enhancement; so long as I regard this negativity as the source of higher positivities that I can create precisely through it, through heroic steadfastness, and so on. [329] I can find myself in the world and still love it because it contains and makes possible much that is worth loving. *Of course, I cannot appreciate it as a perfect world for this reason, as long as I cannot recognize that the ideal of the most perfect world necessarily entails that the world contains chance and fate within itself.* There are here rules regulating ethical behavior and rules of reasonable and self-ethically necessary evaluation of fortune and misfortune, rules of the positive fruitfulness, for the well-being of humanity and for one's own well-being, of misfortune in all forms.

<§8. Ethical life and true happiness. The impossibility of a life of pure joy. World affirmation based on the relative perfection of the world. The necessary practical demand on the world.>

So, finally, how do ethical life and blessedness (satisfaction) relate? Ethical life gives self-satisfaction as a condition for the possibility of each further satisfaction. It gives me value and is recognition of my value or satisfaction concerning my value. The struggle for ethical life is the struggle for me to care for myself. The fulfillment of my duty, however, does not make me happy.

"*To be happy*" can mean: to feel pleasure, to enjoy, be it sensual pleasure or higher pleasure. Naturally, a life in passive pleasure is of low value, though we do not say that feeling sensual pleasure or enjoying sensually is undignified or that the sensual value is not a value. This demands its own value considerations that ethics carries out. In any case, joy and especially joy in fulfillment—the realization of value intentions through one's own deed—is itself a value and a good, both in itself and, relatively speaking, in relation to the demand of the categorical imperative.

The wish to live in joy is legitimate. But the ultimate goal of living in joy all the time is illegitimate, is unreasonable. Not in every sense. After all, I cannot strive for anything without implicitly co-striving for the joy of success. The value for which one strives is the enjoyable, but setting for oneself the goal to savor, to enjoy constantly, is unreasonable since I know in advance that it is unable to be fulfilled. There will certainly be given some dis-joy—if anything is certain. And if adversity, suffering, sadness comes, then I cannot withdraw from it without further ado and, above all, not by turning away my gaze. I may not be permitted to escape it at all; as each case of noble sorrow proves, it is my duty to appreciate it negatively. One must overcome sadness by appreciating it; one averts from that which is mean, <in order>

not to soil oneself with it. Noble joy must also be appreciated, lived through appropriately; and the fact that one lives in joy may also be rejoiced over appropriately.

If happiness is a life of joy, then happiness is a life occurrence but is not the shape of a full life. In in this sense, true happiness is joyfulness in the success of the best that one intends, in the success of the best of all whom one loves and, finally, for all people. One's own happiness is not enough where other people are not happy in the good. Thus, *I can only be properly happy if I do not think of the unhappiness of others*. The satisfaction of my own success is necessary. But the greatness and purity of this joy depends on the joys of the others. And the unhappiness and happiness of others weakens or strengthens my own happiness in the corresponding measure due to the orientation in which they are given to me according to their proximity and remoteness and the degree of the vital interweaving of my own existence and striving with them. This is also necessary and has its teleology that is to be further considered.

I myself have the possibility of bringing foreign suffering nearer to me, to make it intuitable in accordance with its reality and presumability, and, thereby, to restrict my own joys and even to cancel them. It is an ethical question to what extent I should do this in order not to be an egoist. In any case, a value would not be a value for me if I did not rejoice in grasping it. And without joy, the world itself would be worthless. So, happiness belongs to value; that itself is necessary and analytic. But a continuity of success is beyond human power, and a continuity of joy cannot exist for anyone, and even a life of pure joy is impossible for the righteous human being who has a stake in others. [331] A blessed life in this world (and perhaps essentially in any world at all, in any case, any world of this type) as a life of pure and at the same time "greatest" possible, highest possible pure joy (unclouded by suffering) is impossible. And if it were possible in the form of making myself deaf to the suffering of others and to my own, it would be unworthy and bad. But is there not a *greatest possible blessedness* in the highest possible form of joyfulness, and is there not a joyfulness in all suffering, both in one's own and in that of others?

We have already spoken of the *self-satisfaction* of ethical people, of genuine self-satisfaction that consists in the thinkable satisfaction of persisting and having persisted in the struggle against "inclinations." This is a satisfaction that is not a sated enjoyment, but has before it the eternal struggle and the vigilance that precludes a passive dwelling in satisfaction. This is a *reflective joyfulness*. But the human being lives looking ahead, looking directly at their environing world and the tasks arising in it. Life-work in the world is carried out within the framework of the cat<egorical> form of life-tasks in general, gained through reflection. And the world must now, for its part, meet the conditions for the fulfillment of the tasks, even in the passivity of the conditions of experience, passive valence, that is, amenability, so that life, that is already an ethical one, does not move on a lower value level and so that higher value intentions are not constantly shattered, and so on. In accordance with this, I must, on the one hand, demand *from myself—first—an ethical life, and* I must unconditionally comply with this categorical demand. On the other hand, I must also *demand something from the world*, namely, *that it has a shape that accommodates my rational expectation.*

I can be happy in the sense that I can *trust* the world and, at the same time, I can be happy due to a rightfully acquired self-confidence, that I can gain the consciousness that I belong to a human world within the framework of the objective world that enables infinite possibilities of ethical striving and a rational world-formation, a formation in which humanity could, through its own work, progressively arrive at the greatest possible happiness—not the individual human being, but humanity. [332] But also the individual human being, insofar as he learns to feel in solidarity with humanity and can still joyfully affirm their own suffering, death, irrational failure that arises from their accidental nature, and so on, because this does not cancel out the "harmony" of the whole, the value of the whole that increases into endlessness and, even better, carries it along precisely through this affirmative sentiment. *I can only be completely happy, if humanity as a whole can be happy*, and it can only be in this sense. Life is worth living under all circumstances, already as an ethical; it is worth living for the individual who understands their ethical task and takes it upon themselves. It is worth living even more when I, the agent, see an open horizon of social connections of love and community of work in which we can all make average progress and help each other in the elevation of existence. That is already more and gives a higher value to life. And even the expectation that the world is such that ethical=reasonable community can be constituted, even if only for the time being, this already is a relative perfection and creates happiness, namely as world affirmation. On this happiness, however, rests the deep shadow of the irrational barrier. I cannot be satisfied with the world if it is not better. *The ideal* is *that the world is indeed a perfect one*, this is *the necessary ideal or the necessary practical demand that I must posit onto the world*. The world is as perfect as it can be thought of, fundamentally in the sense of the opt<imistic> apprehension that human beings and the world are attuned to one another, that the world has a structure that makes an infinity of ethical culture possible for humanity, in endlessly progressive ascension, out of one's own ethical freedom, and included therein the possibility that ethical sentiment may propagate and become effective from human to human by example and instruction, and, above all, that the endless task of ethicizing humanity may be set and progressively realized. But this is not enough. It would also have to be an essential necessity that contingencies that are incalculable for the agent and even that irrationality of the agent themselves are indispensable for the perfection of the world. And perhaps I can only know this as a fact from the *a priori* essential necessity. [333]

I could be satisfied in the highest sense if science showed to me, if I could evidently intuit *that the world is not only actually so but must essentially be so* that a world-totality and the I-totality included in it, can be necessary only if this world fulfills, not only the theoretical conditions for the possibility of concordant being, but also *conditions for the possibility of concordant value-being and of concordant being as a field of human rational praxis*. But to this belongs, in advance and in itself, that the world is this way and its human beings similarly, that such a *science proceeding into infinity* is capable of developing.

Husserliana 42, Appendix 29: <Self-love and neighborly love. Striving for objective values. Intersubjective world of objects as a condition for the possibility of valuable subjects. God as a teleological principle.>[16]

Congruence of happiness and selflessness = customary morals. The moral as field of "selflessness."

Theoretically, what is valuable in me and in others is the same; and the command to love one's neighbor will bring this to validity. On the other hand, there is no command to self-love. Commands – prohibitions. Therein lies the sense of a "social" response. A demand that the other makes in the name of reason on me as a valuating and acting subject. He says to me: you must not take such a position, otherwise you are unreasonable, otherwise you will arouse my displeasure and that of others.

Motherly love is not commanded to the mother. Moreover: there is no command to satiate oneself, to eat is not commanded, at most, only how much or when to eat, and so on.[17] *Self-love* is self-evident. *Neighborly love* is not rare, but it is commanded, we are reminded of it because we, as a rule, do not allow ourselves to empathize with the interior of others and we sin out of an unjust preference for ourselves. A habitual power of self-preference that, in a certain sense, also is just. In a certain sense, I am my own neighbor. Thus, *self-support is an imperative of ought*, of reason, [334] and of the possible demand that can be made on me by others as representatives of reason. Self-support is natural, but often erroneous. One errs out of sensual passion and prefers the lower sensible value over the spiritual one.

The merely subjective values <are> like the merely subjective objects (data of sensation, sensual feelings). Here there is only brute "concord." Others can have the same sensual data and the same feelings in relation to them.

Objective values. My conviction (my "cognition") is mine, and another can only have another conviction, their own, with the selfsame truth content. But here we necessarily have overarching concordance between all rational beings; and the same truth becomes, in every subjective realization, the same goal and has an identical ideal value. Here my value and my neighbor's value is the same. But by striving for it and actualizing <it> in convictions, he gains, in his personality, subjective values related to the same objective value. *By all striving for objective values, we all gain subjective values* and in such a way that our hearts harmonize with one another and we must, to that extent, mutually love one another. Every objective value <is> a moment of possible socialization.

Striving for objective values makes each one better for themselves, and, at the same time, creates for him a pure satisfaction that is itself a pure value. In ethical deliberation and selection, a subjective life grows such that is "harmonious" from the standpoint of the heart, <an> overarching concordance of hearts in the form of a

[16] <*Husserliana 42*, 333–337.>

[17] But can love be commanded? That is being denied.

constant, purely satisfied, blessed life. Axiological and practical perfection of the subject in their life. Best possible, blessed life. The human being in community. *Objectivity* as a harmonious concordance of all subjectivity in cognition, as harmony of the hearts of all subjectivities, as a condition for the possibility of mutual love, the greatest respect, appreciation, greatest possible peace (least possible hatred), greatest possible mutual exaltation, "support," and greatest possible culture. But is not culture, considered in ideal heights of value, there to elevate "human" subjectivity, to make possible valuable subjectivity of the highest possible value? Objective values are, of course, values "in themselves," just as objective truths are truths in themselves and just as objective objects are objects "in themselves."

But if we carry out an *axiological examination* of the world, we say—and this is the transcendental-axiological consideration: *if an ego is to be a valuable and as valuable as possible,* [335] *it must constitute an object-world,* since such a world, as a world of knowledge, is the condition for the possibility of valuable knowledge of higher and the highest levels (the latter if it makes possible "science" in the exact sense and, thereby, truths in themselves). If subjectivity is to be value<able> in a high degree, there must be many subjects, each of which has the greatest possible value. And the many subjects must constitute a single world, they must be there for each other. Why is that? The question arises whether this <is necessary> for the sake of increasing the cognitional values.[18]

But cognitional values, doxic values, are not the only possible ones. We also have the values of beauty, practical values. Cognition itself is a kind of praxis. Beautiful things, works of art. Practical configurations of objectivity according to ideal ends, according to values in themselves.

The thought of Fichte. A valuable subjectivity demands a world as an objective world and demands it as a field of ideal formations through subjectivity. The world corresponds to the principles of axiological and practical reason (which should include the principles of theoretical reason). A subject is thinkable without a world for the subject. (Many <subjects>: but not without the beginning of objectivity.)

But subjectivity—everyone can state this about themselves—is such, I am such that, for me, the conditions of the possibility of a reasonable life are fulfilled. I can become free, I can become a valuable ego being.

It is as if the world itself were there for me for the sake of its value. And I myself were for the sake of being a valuable I. And since I know others in the world, I may say: in order that we all not only are, but are as a multiplicity of beings that are as perfect as possible, and in order that each of us not only is, but is in the greatest possible perfection, we must not only be, but must be there in multiplicity and must be members of a world formed in this way, namely, a rational world, to which we ourselves lend ever greater rationality. The world will be the best if we make it the best. And this means: the spiritual community will be the best if it, as a self-objectivating and world-constituting community, makes this world axiologically and practically the best.

[18] Many subjects multiply the cognitional values.

What then is *God, the theological principle*? If subjects are thought absolutely in transcendental purity, and if they must be thought as "immortal" while only the objectified subjects are mortal, then the question is what the fact of the rationality of the world means according to its being-in-itself in natural respect and according to its ethical rationality as infinite task of free beings, [336] and what the question of the Why and the Wherefore means.

If we say that God created the world just as we say an artist created his work, then the "intelligibility" of the Why and the Because here is that we understand the action and understand it with respect to its motive: within a subjectivity that is related to an objectivity of which it is conscious. Of course, this is useless here.

Here, the answer can only be given by delving into the fact and the essence of the irrational and rational. *The ego cannot be made* nor can egoic life be made. *An ego can only be by developing*, by becoming in a constant interior motivation and thereby also developing into freedom, then consciously taking a position and thereby being motivated in the manner of awakened reason. Such is each ego. In absolute terms, we have, therefore, the development of "monads." In empirical terms, we have a world in which all egos have an objectified or an objectifiable place. And corresponding to the original inner development of the monads is an exterior counterpart, which is also an index of inner developments: development in the biological sense.

And finally, one is led to ask whether the whole empirical world is not, as in the thought of Leibniz and Brentano, following Aristotle, a system of development, an infinite organism. The interior system of the development of the monads in their mutual "harmony," then, corresponds to this. In the transfer of the Aristotelian entelechy theory into the transcendental, the thought Leibniz attempted under the title of "confusion" presents itself. This would mean that all of our hyletic data is already a "product of development," and thus has a hidden intentionality that refers back to a "synthesis." Everything leads back to a πρῶτη ὕλη, which is fully undifferentiated matter, and to primordial constitutive processes with intentional motivations. However, through all development run leading "ideas." The final leading idea is the total idea of overall development. This would be the system of ideas of formal "mathesis" in the very broadest sense (spanning axiological and practical, and so on). The system of principles and the ideal unity of all formation.

Development, however, is not to be seen as a beginning and ending process, as a sort of little history, but rather the first ὕλη is always also there and the ὕλη is in everything developed; conscious egoicity is also constantly changing into dying, that is, everything sinks in the form of passivities (loss of memory, and so on). The main thing is: to ask about the grounds of a development, to pose the uniquely rational question, and to describe the development according to its deepest sense and to describe its leading ideal goal and its structure. God is the entelechy and outside Him there is "nothing"; [337] He is the all-forming, and the irrational matter is not a made thing, but mere stuff. And the world has its being from God and is otherwise "nothing." And God is only as a leading and "animating" principle of perfection, and so on.

******

That does not take into account the following. Values in my own perceptual sphere, my sphere of direct experience, come to original perceptual self-realization; I savor them myself. Values in the perceptual sphere of others, those that they can savor—these I do not myself originally savor. This gives the egoistic striving a power that the altruistic one does not initially have.

It can now be asked: is there not in itself a greater value of my subjective values than that of alien subjective values? If it were a matter of sensory things, you could remain with that. Why should it be better, or even equally good, for good taste to be realized in the other rather than in me? The matter is different when I examine the values of absolute ought; and these are specifically personal values. In relation to these, sensory <values> are merely that which underly and are means for possible personal elevation, for preserving their natural sources of power and the like. Of course, the fulfillment of my personal absolute ought is also only an original experience for me and so is preferred. But here the law of love rules. As soon as I have the other given empathetically as the subject of their personal ought (even if not originally experiencing them myself, at least actually intuitively experienced), I "must" sympathize with the other, love them. And to support the other as this subject is my absolute ought. The values of neighborly love are by far the largest part of the values of an absolute ought. One can say, of course, that for a solipsistic ego there are already absolute values, values of "pure beauty" (including even purely sensual beauty), values of beautiful art and values of knowledge. But even these increase enormously as soon as they are constituted as personal communal values and establish, as communal purposes on which people are working collectively in the absolute ought, noble communion between people.

One can say: in principle, all absolute values are values for everyone and not merely subjective. Only the means for absolute values can belong as exclusive property to individuals. But this has to be considered in more detail. [338] Naturally, this is not simply to say that loving support of others in the absolute is better than one's own, in itself, though it is against the neighbor that most sins are committed.

Husserliana 42, Appendix 30: Self-preservation. <Idea of an ego that strives for and realizes its true self in the best possible manner.>[19]

<Content:> The ego (the psychic subject), the community as well, as preserving themselves through passive and active activity, through external defense, through active appropriation of external aid, through external achievements and through achievements in themselves.

Preserving itself—not as an identical, rigid unity of properties, not preserving itself as a thing preserves itself, be it in remaining unchanged in its states, be it in

[19] <*Husserliana 42*, 338–340.>

preserving, through strictly causal lawfulness, its properties while changing in its states.

An ego has no possible fixed properties. It is not a unity of change in the sense of a physical substance; it does not have a substantial being in this sense. It is rather a unity through a tendency towards self-preservation; it is by striving to be and striving to be itself. It is itself an idea, either an idea lying in the finite, such as that of the individual animal of the type "lived-body," or an idea lying in the infinite, such as that of the individual human being who is, and is as this human being, by taking on tasks and either fulfilling or failing them, and sets themselves tasks that are either their true tasks through which they could realize themselves in accordance with the sense of the idea, or false tasks that move them away from the idea. The human being's true self is that of a value idea that continues to be realized in the setting of tasks and the solution of tasks. And this idea is itself an ideal to which the striving life of human beings comes more or less close.

Virtue is the habitual direction towards this ideal, specifically, human self-preservation and, more precisely, the preservation of the individual, their own self. The typology that prevails in this respect in human beings in general, the general essence that connects all humans and brings general virtues with it, [339] is the condition for a possible general doctrine of virtue for humans: the idea of a self-preserving ego in the "spiritual sense" as such.

"To preserve oneself": to preserve the true self to which each human ego is "directed." But why preserve and not, say, arbitrarily form and realize? If we speak of preservation, we already assume that a human being, a normal human being, "wants to go toward" their true self, that they are already "directed" to it. We assume that their education, which they have either received from others or by self-education, has already directed them to it; we assume that the human being has already grasped the idea of their true self or that we, their spectators, examine them from the point of view and goal of this idea and examine them as if they were striving for this idea, or think that they should strive towards it. We then have in mind that every striving is directed toward this idea insofar as it is in the sense of the striving to "want" to fulfill oneself completely, and that every fully fulfilled striving terminates in a value, the having of which, as having a true practical value, serves the "self." The true life, the life under the idea of the true self, would then be a life that constantly sets true goals for itself and progresses from true achievement to true achievement. That is, of course, the idea of a divine life.

If we instead take the idea of a life that is inhibited from realizing true goals, insofar as its power is "finite," insofar as "irrational" resistances prevent it from having any effect, we have to reshape the idea in such a way that any sublation of the true goal positing in failure can bring forth new true goals in adaptation to the factual state of affairs co-created by the failure itself. Progress from true goal positing to an ever-new true goal positing, and progress from deed to deed, from successful ones to successful ones or even to unsuccessful ones—"with the effort of all his powers." The energy of the tension is itself a possible "secondary" goal, a means lying on the side of the subject. We now have the idea of an I that strives for and realizes its true I in the best possible way. But the I as a finite I is also an erring I; it can limit the error, theoretical and axiological, by the indirect means of self-education.

Finally: The I is sinful I. It is an I whose activity unfolds against the background of a passivity, a realm of affection, that is not, from the outset and throughout, a realm of rational affection growing out of rational motives through sedimentation or to be rationally legitimated afterwards. Naturally, every I as such has inclinations and has its background of affection, but it is not necessary that every I is subject to the power of blind affections and even those affections that have been recognized as irrational. [340] There arises the idea of a sinful I that seeks its true self and lives the best possible life in the direction of this true self and therein lives a virtuous life in sinfulness and humility.

Husserliana 42, Appendix 31: Custom. <Custom in the narrower or traditional sense. Law, custom in the proper sense. The point of view of custom as "their" point of view.>[20]

The system of general demands of sociality (of the communal environing world) with regard to the conduct of the community members and sub-groups, inasmuch as it is in their power of will (good custom, the ought, the required, bad <custom>, what ought not be, what should be censured).

This includes 1) *custom in the narrower sense* of traditional custom. The model of rule-forms of manners of behavior that each subject demands from every other subject and from themselves, and demands in such rule-forms because of the tradition and as habitually practiced by everyone (themselves included), as having become second nature; and indeed, when actuating the rehearsed manners of behavior with such rule-forms, demands without ever meaning or having any necessity to mean any goal in the <proper> sense (that is, without considering "that" and "whether"), a meaning whereby exactly this form and no other one should be valid. That which is proper to custom is practiced as "form" (social form) because it is handed down as traditionally demanded and approved as such, and for no other reason (perhaps because the sense has been "lost" or it can still be felt, but is no longer a motive).

2) What about the *legal* in contrast to the customary? The legal is also imposed as an ought from "outside"; it bears the character of externality, of coming from the community as demands, authorizations, and so on, but, in its content, it is rationally purposive. It is not authorized merely because it is tradition in this rule-form and is traditionally authorized, but it is authorized and "demanded" as a demand of a commanding will endowed with the power to be enforced (king, legislature). The legal has its rule-form thanks to this commanding and executive will, [341] under which

[20] <*Husserliana 42*, 340–343. From 1920.>

the will of the community member is bent and <the will> of the "authentic" community member knows itself bent in the form of conscious and deliberate subordination.

On the other hand, the law—even if not entirely, but to a varying extent—is at the same time content of good custom. It belongs to good custom not to violate the law (in the good society, not going to jail is a part of it; but to smuggle, that corresponds to custom, it is at least the *Adiaphore* of custom). Different social classes have different customary "moral" attitudes towards the law; different strata of the law, although the same for all, belong for various social groups to good customs in different groups because this very good custom is different for the different classes.

3) *Authentic "customary morality"* according to current usage, ethical behavior, ethical virtue. It is similar here to law. The ethical is the realm of the rationally purposive in the absolute sense. And it again supplies the components of good custom and bad custom. It demands that certain ethical components belong to second nature and to tradition: "One," a decent civilized human being, does not abandon a friend, helps the poor, and so on. Heroism is not required. The morally "correct" is the domain of the moral, which, at the same time, has its "sanction" in custom.

Had association psychology—also in the more developed form of Mill—had an eye for the essence of the pure phenomena of consciousness, it could not have come up with the absurd case of identifying the *selflessness* or *self-forgetfulness* that characterizes specifically moral behavior with that totally different self-forgetfulness or selflessness that supposedly arises from the fact that the egoistic purpose of philanthropy is pushed from consciousness and forgotten, and now remains now as supposedly "selfless," as an end itself. Not only that—as in the parallel example of avarice—the derivation is false; it is, as for Mill, a phenomenon that overlooks the proper sense, the whole proper essential content of virtue. Even if one were to take as parallel examples those of training, be it dog training or educational training, that is to say, if one put the natural development of virtues on the same level as the technical training of children in general custom, in the use of knife and fork, posture, and so on, something essential would be completely overlooked.

Action in the sense of custom, to which we have been trained repeatedly from childhood and onwards in society, has, for everyone who act in such a way, *the phenomenological character of what is imposed from outside, what is required*, [342] of what should be in the sense that deviations from it are frowned upon, are perceived by onlookers as unpopular; even if they have become "mechanical," they also have an intentional sense that one may question, that one—as thus acting and habitually operating—can clarify, can make it intelligible, even if nothing of reason appears in it.

A virtuous action, however, has the character of an ought in common with the act of merely traditional custom, insofar as it is also, as a rule, praised by the outside world, by onlookers, by the society that takes notice or—like vice— blamed, and in such a way that this manner of judging functions as communal and mutual. The agent knows themselves to be praised or blamed by a normal onlooker, as they would praise and blame others in such a case. They assess themselves through the eyes of others. Each social ought has a peculiar universality; each I who feels an ought feels connected through the ought with a community, each member of which would demand the same ought from everyone else and from themselves. Here, too, if one asks about this character of the ought, one is referred back to possible surroundings, to a community connected in mutual valuation and value assessment, in mutual approval and disapproval. *On this side*, however, all general morality, all activity of virtue, all examined moral sentiment, is designated *only from the point of view of custom* and is itself a mere field of custom. One tends to distinguish *custom* and *customary morality*. But if one does this, one retains in morality the moment of custom to which the word specifically refers. What has been described so far is by and large the phenomenological characteristic of *custom as custom*. It is characterized by the *social ought*, in the sense that sociality is thought of as *social subjectivity*, as a connected community of evaluating, of approving, and disapproving persons; and the objects of this approval are again the people, their sentiments, their actions, their motives, their ends.

The point of view of custom is that of asking (loudly or internally and quietly) ourselves as we act: *What would they say about it, how would they judge it?* And it does not matter at first whether you think of this "they" as the general mob of civilization, as refined people, or as a more general "they," and, finally, perhaps as "they" in a more ideal sense, perhaps even as a noble human community that is represented in a noble image of humans through the benevolent eye of a revered human being. Note that this change in "them," this relation to an indeterminate community of judgment in which one includes oneself and from which, [343] if one refuses the ought, <one> consciously excludes oneself, and to <a community> that can be characterized as changing, is not something that merely comes to mind, but rather belongs to every social ought in the form of that which is characterized *as customarily moral*, <as> demanded by customary morals, and belongs to it as an intentional character that can be questioned; in doing so, one sees oneself pointed to possible life companions, to others or to the good society and such more or less indeterminate communities whose kind of indeterminateness is itself pre-figured.

The customary ought is, however, only the outer garment, so to speak, of ethical virtue and is suitable, in an adjusted and living manner, only to that virtue which belongs from the outset to general custom.[21] Where this is not the case, the social character of the ought is, so to speak, a reproduced dress, one that has been habitually passed on.

In no case, however, does the customary make up the proper essence of virtue (namely, if we inquire into their sense). Where virtuous custom distinguishes itself within custom as customary morality, the specific coloring of customary morality and the special structure of the co-judging audience, the ethical chorus, so to speak,

[21] We contrast mere conventional, traditional custom with morality as custom, namely as the inventory of ethical rule-forms that, in the general consciousness, have traditional character and are conventional so long as "everyone," "people" as a rule disapprove of deviations from such manners of behaving and approve of the regular as self-evident, and with the consciousness: this is the way it has always been. And in the case of deviations: that way has never been approved of; or, even: such a thing is unheard of.

is based on the inner character of virtue that contains nothing of the customary, nothing of the social chorus of judgment. What is the proper character of virtue, more precisely of every virtuous action, every ethical decision, direction of will and, finally, of sentiment?

Husserliana 42, Appendix 32: <Absolute obligation and absolute love.>[22]

<On> Scheler's preferring without comparison:

"In this preferring, some content is grasped as the best possible, and a value response is directed at it as the best one. I can love someone with the evidence that I could never love anyone else in the same way—[343] or be enthusiastic about something beautiful with the evidence that nothing more beautiful could exist that would make me more enthusiastic."[23]

This "preferring without comparison" is, of course, still explained here by comparison. But what is indicated is a certain evident (or intended) absoluteness that has its analogue in the absolute ought. For example, a mother who saves her child is absolutely certain, without any categorical imperative that she ought to do this and nothing else; and likewise, someone may be certain that he loves somebody in a unique way, whereby, above all, comparison in love contains a certain absoluteness. One can say here: love is not a momentary act, but a permanent decision for a person. I can decide for several people in this way, and so I can (in accordance with my life horizon, which does not exclude subsequent abandonment) simultaneously love several people. There is, however in contrast to these loves, one "unique love" that seeks and wants the same kind of love to be requited—a full becoming one with the beloved or a knowing unity, a "taking him up wholly into one's own ego—and one's ego wholly united with the other," <a love> that excludes any comparison precisely for the sake of this wholeness. I have wholly "devoted myself," and I can only do that for the one. Of course, there is a problem here. One can wholly lose oneself and can freely give oneself; and it can be asked whether one is allowed to do so, whether it is right, a right love, or whether one may only completely give oneself to God in such love and then give oneself forever, and so on.

There is also the question whether there is an absolute practical ought without any comparison, and so on.

[22] < *Husserliana 42*, 343–344.>

[23] <Dietrich von Hildebrand, "Die Idee der sittlichen Handlung, " *Jahrbuch für Philosophie und phänomenologische Forschung*, Band III (1916), page 228, note.>

Husserliana 42, Appendix 33: Pleasure value—absolute value—absolute ought[24]

<Content:> The initial question of the critique of cognitive action, specifically with regard to its cognitive goal, will soon be allowed to fall away in favor of the general question: How are final goals legitimated? And this leads to the only question that is treated: Are all values pleasure values? And from there: How does pleasure value relate <to> absolute value and absolute ought? And how does it relate with absorption? [345]

We are now, looked at more closely, in a critique of successful actions at a higher stage. I thematize myself with regard to my cognitive actions, possibly with regard to other actions. I consider my "I do this, "I did that," "I resolve to do this," and so on. I want to legitimate it. (Which is a new action.) I behold the goals, whether I have actually reached them. I suppose the action and the accomplished realization to see whether the resulting reality of the deed is indeed in accord with that at which it was aimed—for which I did not have to draw on my ego thematically. But I can also *critique the goals* myself—how their values are, what they mean to me, whether I want them as a mere means for other ends or as ends in themselves, and what they are or whether they hold as self-values for me. But what can "self-value" mean here? It can be (and this is the object of the consideration) that these objects <or> occurrences arouse my feeling; that I, while presenting them as realities of my surroundings, also present them as bringing pleasure; that I am affected and brought to action by this presented pleasurable objectivity, which I know as practical possibility, that I am carried away, more or less "passionately" drawn to them, drawn to their realization as bringing pleasure. Also, the activities that I do, productions and their products, can appeal to me in this sense, can be striven for by me, wanted and realized in activities of a higher level as objects of delight. It delights me to make models, music, ski, and so I strive for it and do accordingly; it even <delights me> to allow myself to be moved by the tragic representation in a drama or to hear a symphony, and so on. I do what brings me pleasure and realize pleasing as pleasing. This also includes pleasing ego-activities, actions.

But it can also be otherwise: the affection that determines me to act or possibly to act (to the realization of presented actions) need not be a pleasing one. I take up the New Testament, for example, to "edify myself," not to enjoy something pleasing and to get myself into a "pleasant mood"; to give myself to the pleasing in such passivity of feeling, may the pleasure be a "higher" or "lower" pleasure, the pleasure value of this or that "rank." The words of Christ becoming living in the action of "deepening"—in the action of intuitively realizing their meaning—but they are not "beautiful" as pleasing things are beautiful; they do not appear before my soul as pleasant things with which I could "amuse" myself in some sense, no matter how high. They seize me, inspire me, exalt me. They may admonish me, make me feel my unworthiness, but this admonishment, [346] like that of any revered friend or

---

24 <*Husserliana 42*, 344–347.>

teacher, is an *admonishment of love* that speaks to me in the "most interior part of the soul" and awakens love as an act of love in me or, from the start, moves the ego in the medium of my love and the loving exchange that moves me. Here, love, the loving surrender, becoming lovingly determined and determining or suffering, flows out of and into the ego in a special manner.

Every love value, *each loving pleasing is fundamentally something other than a pleasing of pleasure*, other than a giving in to pleasure or, rather, being-lost in it. The lover does not lose themselves in love, but lives in a special, elevated way as the ego in the beloved, who is the beloved individually for them, from whom they receive beauty as their beloved and a beauty that beautifies themselves as the subject of love. This is so with every pure and genuine love. *From the value of love emerges the loving deed.* A beloved or showing love to a beloved and thereby beautifying from proper sources of love is something other than realization a pleasure.

Values of pure love—absolute values for the person. Values of pure hatred—absolutely negative values, disvalues for the person—not detestable in the sense of pleasure, but rather a disvalue that makes the ego "unblessed." *Absolute values* in the practical realm are the practical absolute ought, positive or negative ought. *Pleasure values values of utility*) weigh against each other and differ in terms of better and best. Every absolute value is absolutely good. But in the practical sphere, one absolute value can be "sacrificed" to another. *Absolute or authentic values do not stand under a law of absorption.* Pleasure values are just pleasure values as long as a greater pleasure value does not compete with them. The absolute value remains what it is. A (compatible, not detrimental to one another) sum of pleasure is a "higher" pleasure value. A sum of absolute values—how do things stand there? I am also richer in love when both are. Only that the value of the sum does not absorb the value of one alone, but rather the absolute value remains value. But how about practically? Where both cannot be realized, where the reality and realization of the one cancels that of the other, the *absolute excellence* of the one may emerge. What does this mean? The realization of an absolute value is an apodictic activity of willing—in the cases where its disclosure is not apodictically required. Blessedness can be inhibited, tarnished, even abolished through an unblessedness. This is so where an absolute value and an absolute disvalue are united. *The disclosure of an absolute value* yields *an evil*, since the non-being of such <values> is an evil. But the disclosure of an absolute value, where it practically competes with another absolute value and must be chosen, can be an "absolute demand," an absolute ought. [347] Therein also lies an absolute value. Every apodictic will is an absolute value, correlatively to the deciding for the other absolute value. The preference for A over the sacrifice of the B is the ought. So, it is not that A *per se* is obliged just as little as B *per se* is not obliged (in and for itself it would be obliged), but, in the choice, one practical ought absorbs the other. Only the one is apodictically required. The absorption essential to the absolute ought determines the higher value. Or are the values as values in themselves higher? Is this love a "higher" one? The joy in the representation of actual being <is> a greater, a higher one? Mother—child, mother—fatherland. The beloved—the beloved fatherland. One can say that there are already relations of values in love itself, despite the fact that I had reservations and decided otherwise.

My concern: Can I say that it is better for the child to perish in body and in soul than for the fatherland to deteriorate? Can one seriously make such comparisons? On the other hand, one would like to say: without comparing in this way—according to the magnitude of value, the scales of value—I am still, in some cases, obligated to *an absolute ought in the choice. I disclose what is incomparably valuable to me because I follow the ought in my choice.*

Chapter 2
Husserliana 42, Text No. 25: <Value decisions and the call that goes out to the ego. Value conflicts and value sacrifice.>

Edmund Husserl

<§1. Acts as actions: Experiential actions and predicative judicative actions. Valuing acts and the call of true values.>

Science. Judgment, insight as self-possession and insight into necessities and universalities. The Why and the That. Explanatory science. The causal explanation (Why-explanation). The various Why. What is to be expected, under given circumstances: as necessarily co-existing. Necessities of fact as fact, empirical laws (laws of fact) and *a priori* necessities. Necessities to have to judge in this way, necessities of consequence and necessities of right judgment, of judgment that is directed to the being grasped in itself as itself, immediately or mediately. Correlatively: necessities of being. Being in the broadest sense. Necessities of matters. Material being versus being of value and being of practical ought.

1) *Experiential actions* in the broadest sense, also actions that aim at the production of original anticipations, to the possible and conjectural prior to the bringing out of the actual, that is, prior to the (not always feasible or immediately feasible) execution of the experiences that make the individual visible according to its intended sides, and so on.[1]

[1] "Experience": individual doxic intentionality. The experiential explicating, the taking together, the experiential following of indications (inference), pictorial presentation (pictorial indications, mirror).

This chapter was originally published as Text No. 25 in *Husserliana 42*, 348–355 dating from the 1920s.

E. Husserl
Freiburg im Breisgau, Germany

S. Heinämaa, A. J. Steinbock (eds.), *Values of Love and Ethical Reflection*, Husserlian Legacies: Themes for the 21st Century,
https://doi.org/10.1007/978-3-031-68698-6_2

2) *Predicative judicative actions.* Conceptual thinking and knowing: thinking individuals under universal concepts, subsuming individual states of affairs, judgment complexes and possibility complexes, [349] under rule and law, "inferring"—always in the purely predicative sphere—or ascending from individual experiential complexes through conceptual thinking to general judicative complexes, to essential complexes and to empirically universal complexes. Thus the "ascending," ἐπαγωγή, in both senses: of ideation and of empirical induction. On the other hand, thinking in the sphere of pure *Logos*, of logically grasped statements. Also included in this, signifying spoken expression and linguistic thought.

"Presentation" and its variations (individual intuition). Connection of presentations, or of the presented. Judgment and its variations. Connection of objects of judgment. Everything in actions.

Intentionality, intentional horizons—horizons for possible realizations, realizing actions.

Presentation (experience) as doxic consciousness of the individual—on the other hand, presentation as the iterable function of self-giving.

The judging and its derivatives as "presentations." Presented states of affairs, self-given or not self-given. Judging about them as substrates, and so on.

Acts of liking and disliking, of "valuing." Valuing activities. The "experiencing" directing oneself to value and disvalue. Valuing as a universal function.

Attaining individual matters and material connections, grasping them in themselves; the attainment of predicative states of affairs, of predicative logical formations, of laws, of explanatory determinations from laws—grasping them in themselves. The attainment of values and value complexes and of predicative "truths" or well-founded anticipatory convictions about values.

The action as a universal function and the action as the generation of material formations, then of truths about things, of values and value truths, and so on. Action–realization. "Motive" of the action of the ego. What determines me as an agent. Affection—in the drive. I am driven, I follow the drive, the pull. Empty affection? Drive toward a "presented" goal. "Sensory" affection, affection by higher, lasting values. Affection by truths, by "true values." Presented truth as realizing end. [350] The call of matters themselves, of true ones. The call of *true values*. The call of practical values and of the practically best.

<§2. General objective values and absolute subjective values. The essential differences of practical ought: the ought according to the rule of the practically best and the ought according to individual values of love.>

The general values and the individual values that are of worth for the sake of their generality. The rational that has its general Why, and its Why that finds its answer in generality. Yes, because one moves only in the framework of generality or rather of indefinite possibilities. It is otherwise in the sphere of the real. The real always remains in part irrational according to its existence. The "irrational" existences and values, the irrational truths (truths of being). The irrational truths of value and practical truths.[2] The irrational and its rational form. The irrational "material," the insoluble remainder of the this-there. The analogue of the this-there of the material sphere in the value sphere: the individual value in its rational form and its insolubility.[3]

The essential differences of the practical ought:

1) The ought that belongs to the respective *practically best* for me, or the ought that belongs to goods as objective values and to the best as the objectively most valuable among that which is realizable, between which the choice hovers, whereas each has its own weight in the choice. That which is useful in the lower and <in the> highest sense.

2) A different kind of ought. In the case of the previous ought, the violation of the rule of the best is practically "unreasonable." One gets too little, one remains in the enjoyment of what is less valuable. Can all practical aims be regarded as practical values in a choice; as something that is possibly weighted above something that is weighed lower inferior factors? [351] What makes the difference? –1) Intersubjective values are those that can appeal to anyone depending on circumstances; <2)> *subjective values (specifically egoic), exclusively addressed to a determinant ego.* But is not value as such something that everyone can "have" in the corresponding valuing and that everyone can determine?

ad 1) *The valuing of the "uninvolved onlooker"* and the values that he can "really" understand accordingly. Values that anyone involved can have in original intuition.

<ad> 2) The motherly love and the child as a practical value for the mother. Can just anyone really and authentically understand this practical value accordingly? Can just anyone understand the practical ought that addresses the mother as mother? If that person too is a mother, then she can understand accordingly by analogy, but only in the following sense: what my child means to me, her child means to her. Every mother can only experience in an original intuition her maternal ought in relation to her child.[4] On the other hand, anyone can experience the objective value of the child as an "uninvolved onlooker," whether they are a father or mother or whoever. As such, one has a liking (or disliking) and, with respect to the child, a pleasing through value-apprehension and value-beholding. This experience is a rational motive (or contains a rational motive as value in itself). In the choice and where self-beheld values are competing, each has their weight, and the best is decided by the achievable maximum weight, and, in the case of equality, it is decided

[2] Why irrational? Merely because they concern the this-there?

[3] But is there not a double sense of irrationality in the value sphere and practical sphere of that which is not comprehensible from its generality, from its general Why?

[4] Thus, "child" is an object that, in addition to its universally accessible objective meaning, has a layer of meaning that *only particular egos can originally experience "in principle."* Thus, *not every objective* is *intersubjectively experienceable*, realizable as itself by experience.

by choosing. But the mother does not choose, and determining for her is not the value that anyone can value accordingly.

The question here is: How does the practical ought relate to the valuing in such a case? Every ought is addressed to the respective ego that ought to do something. How would it be if we said: The ought naturally has its "value" for the ego of the ought? *But there are values that are attached to the object, to its uniqueness*, that affect the respective ego in a valuative way, that is, excite the ego's liking. Or, practically changeable objects show, when their possible changes are made intuitive, [352] a change that would create an increased value and, on account of it, one for which is practically wished. The ego is *valuatively affected* by the possible object, and by then anticipating or beholding the object-based value, the ego is *practically affected* to realize it.

But in the other case, it is not the character of the object and its generic value that practically motivates and prompts the practical ought. Nor does the τόδε τι of the object come into practical question as a mere bearer of its generic kind <and> of this value which is generically motivated, thus not merely as being so established and to be so valued on the basis of this establishment.[5] Rather, *from the ego, a loving value streams onto this individual object* (which, of course, has its individually specific character and accordingly its objective value) and gives the object a value that, as a value originating from this ego, is from this ego and is special for this ego. And this value is followed by a practical ought that *the object-based value does not determine* (the one originating from the object by affection) and that *stands outside of all value-relations with the possible object-based values of this and that object and, as practical ought, does not stand in any choice in which another material value besides the loving value "might" have weight.*

Within the class of absolute values for me, for this ego, values that originate from myself (even if the same holds for all), *conflicts of value* can then arise as well as *sacrifices of value*.[6] But it is a question whether there are here, in the specifically individual sphere, only differences in weights and how the special forms of essential laws are to be described, for example, <the> duty of the mother to her child. [353] She has other duties quite individually as a person, for example, the duty to the nation or to the family or to an adopted child or to personal friends. All such values and duties rank for themselves on one level, but in the identical reference to the ego that is here value-creative for herself.

---

[5] So, the object as an individual of its kind is bearer of value, and this is from the beginning and pre-logically a difference of the mode of consciousness, and certainly a difference of the valuating consciousness.

[6] Tragedy of personal conflicts: I should not simply renounce a "good," abstain from a delight; but if I decide for one "good" when several are incompatible, then I sin against an absolute ought. The preference may be "evident" and absolutely obliged, and yet, I must sin, I must do something for the sake of higher right that is nevertheless wrong for me. Thus, in the individual value sphere. Where the vocational activity proceeds upon intersubjectively identical values, it is more advantageous.

Vocational duties (in the higher sense), value achievements of a kind for which I have decided from within. Religious duties—do they also belong here?[7]

Individual values go from I to another I or from the I to itself. So, they are self-relative or alien-relative. On the other hand, persons can be intersubjective values, object-based values, insofar as they belong to the intersubjective world, are comparable with each other according to values and capable of standing with their demand in practical choice for any acting person. Cannot other objects in the sense indicated also have individual value, value for a determinate ego, and even as unconditional value for it?[8]

What about works of an artist and for himself? Objectively, they have aesthetic value. For the artist himself, they have emerged from his artistic vocational will and have individual meaning for him. They belong to the circle of his *absolute responsibilities*, his absolute self-responsibilities and responsibilities before God and all appointed judges. Thus, among such values of individual, personal meaning there is a group of its own. Vocational creation of object-based values and—as in science—work on objective value systems—these deeds are precisely deeds directed toward such values out of an *absolute ought.* [354] The broadest concept of vocation coincides with the concept of this absolute personally originated ought. Indeed, one also speaks of the maternal vocation.

What about the neighborly love, the "*duties of love*"? The pure and *genuine love from person to person.* The value-appraisal of another according to their spiritual beauty, their education, their vocational competence, and so on, is not yet love, even though it is an appraisal, delight in the other.

Love is the delight that addresses itself from my interiority as this I to the other, in which this other <I> gains a unique value for me and in general keeps it habitually. Personal valuation is thus ambiguous. Each valuation is performed by the I, but only the loving one is primordially rooted in the I as who I am and, at the same time, rooted in the counter-I in another, correlative manner. Mutual love: a union in love, a loving spiritual embracing and connectedness, being "determined" for each other in the consciousness and in conscious habitus, being called to each other in existence, in I-being, that is, in personal life of the sort that the other's life, all their doing and letting be, is personally significant for me, concerns me personally, enters into my personal life and vice versa.

But is not this love as my personal *actus* and habitus something that concerns me and only special others, namely, my special beloved? But then again, can one not say: In general, I gain access to alien interiority only exceptionally; only rarely does it come alive for me and awaken my "personal" interest; only rarely am I open to

[7] What gives the person the high value of dignity is following their vocation and living according to it. Thus, persons have objective value primarily by following their individual ought. Individual values, values for an ego as ethical person, are thus decisive for the true objective evaluation of persons, namely, depending on whether they make the singular personal good (to which they are called) their life's task.

[8] If we conceive of an ego that alone experiences colors in the world, then values of colors are accessible exclusively to them. But they are not thereby *essential values for this ego*, not their vocational values, not personally originated, not "personal values" in a certain sense.

such "I-calls"? But I can become so and I can do something to become so. Of course, there remains the difference between the love that, like a motherly love, originally remains there and remains distinguished in its personal stance, and the wider love of human beings, as already in friendship, which is also individually distinguished, though it is questionable to what extent it is uniquely distinguished.

In any case, love is related to the person individually, but thereby at once related to the value of the person. In love, there is an evidence of "valuable and dear to me." But this evidence has its indeterminate horizons. If I ask what I love about him (insofar as I love him precisely as who he is, as ego), then these are properties of value and properties of the development of value. [355] Every protruding disvalue violates my personal love and is itself not co-loved. Neighborly love as love of the other, insofar as he is good or can become good, insofar as he strives toward the good, even if it is vague, and, rightly guided, could strive toward it.

Duty of the love of neighbor: the practical ought that springs from the love of the true ego that becomes and is in the other or the possibly becoming in the other, the duty to support them wherever possible, to strengthen them by expressing love and joy, their becoming-good and doing-good, and so on.

The sphere of personally rooted values as the sphere of general norms. It belongs to the "essence" of each personal individual, each I, that they have their sphere of personal decisions, personal love, personal ought. Of course, the talk here is of one's own essence, for the I does not have an essence like any matter. –Types, classes of personal values and corresponding general and *a priori* norms.

Naturally, my practical sphere is altogether my own, uniquely belonging to me; this holds both <for> my objectively material best of my best possible rational enjoyment and <also for> my wholly personal ought, my categorically demanded. The former would be the same for anyone in my external situation. The latter is *from the outset only for me*, insofar as it arises *from myself* in relation to my personal environing world. But even if no other ego can personally evaluate as I can, since otherwise they would be me, they can personally value *in the same form* (in the same type, though in relation to other objects). Concerning neighborly love, the other can also personally relate to the same correlate persons. But only from a personal call. The universal type of the personal forms of vocation and love determines universal norms.

Husserliana 42, Appendix 34: <Two values: objective values as absorbing values and love values as absolutely subjective values. The decision for values of love follows an individual call of absolute ought.>[9]

There are here eidetically regulated specificities of the values that differ in their origin and contrast with one another in accordance with the polarity of the direction of their origin. These specificities are expressed in the following: With one sort of

[9] <*Husserliana 42*, 356–359.>

values, I have a choice, only that the choice should be a reasonable one, I should choose the best among the practical "goods." In the case of the values that receive their personal sense from the depths of the personality and its personal love, there is no choice and no "quantitative" differences, namely, no differences in weight, outweighing and outweighed. A value that springs out from myself, for which I decide on original loving devotion as the one who I am, is practically considered an unconditional one, an absolute ought, binding me as who I am. To decide against it is to be unfaithful to oneself, to lose oneself, to sin, to betray one's true self, to act contrary to one's true being (absolute practical contradiction). Whoever decides unreasonably (without understanding, impractically) is unwise, stupid, and damages their value in a certain manner. Perhaps we express ourselves in this way: he damages their personal value. But strictly speaking, this expression has generally a deeper sense. It is the specifically personal value that gives dignity to the personality.

A personality has value, personal value, in double direction:

1) in the objective direction, in which a personality evaluates and practically selects the best objective values and thereby expands the realm of objective values in the world;

2) in the ego-centered direction, insofar as a personality decides, as a personal individual, individually for what they grasp lovingly as this ego, grasp as something of value for itself, as something that in the realm of values is most originally its own, as a wholly personal matter and concern.

In a choice between merely objective values, I also still have something such as a categorical imperative and a kind of absorption. To choose the lesser good and not the best overall is "unconditionally reprehensible"; the lesser good is no longer a practical value, it is crossed out as a practical value. But nevertheless, it still "carries weight" and has its right as a value, [357] but only as one that has been outweighed by the other. In contrast to a value that is absolutely rooted in the I itself and that springs from love (as absolute love), an objective value itself carries no weight, unless this objective value is embraced by the absolutely valuing I and belongs to its goal content. This is meant as follows: the mother is under the absolute ought of her motherly love to care for her beloved. But in this intentional direction of caring for the child lies supporting the growth of the child in the direction of the values of the child's personal development, and indeed in such a way that makes a human being valuable and this human being here valuable (that is, according to all kinds of abilities, but also in the direction of the child's specifically personal action proceeding from its personal will). Of course, everyone has the task of providing this support wherever this child steps into their practical sphere and whenever such supporting is practically possible; but for the mother, the practical good of this child has even wider value and absolute subjective value. Within her absolute ought ("Support your child!"), value comparisons and choices naturally then play a legitimate role. But only on the subsoil of the absolute subjective value form "my child"!

Many phenomenological questions remain that need to be considered individually: all values affect, and all original valuing is a decision by the I for the value. To stand in a choice is to "isolate" each value individually, to let oneself be guided by the affection toward each value in a "responding" valuing comportment, a comportment of evaluatingly deciding for what exercises the affection. It is precisely therein that the value is actually experienced by value-apprehension, is seen and captured in its presentation. But this isolating decision, which as original value-positing, value-determination, awards value, becomes now foundational for a comparative analysis, after suspending the isolation; and one may become conscious of one value as higher, others as lower, and so on. From a comprehensive decision in the isolation issues a valuing comportment of a higher level in which one becomes evaluatingly aware of "greater and lesser value," and gains a preferential value decision for the more preferred and the most preferred. If the values at issue are practical ones (each of which I "can" realize), it may be that the value affection (the inclination) determines me to make a practical decision for the lesser value. Even in the mere comparison of values, an inclination for a value, its allure, can blind me and prevent me from achieving the preference as issuing purely from my valuing performance. The decision must be a free one, and it is a free one, when I do not allow myself to be tempted, do not yield passively but actively constitute the value from out of the I, fulfill the value intentions and actively link them into unity. [358]

In general, it is first of all evident here that the self-giving, self-constituting valuing issues from the I-pole and, in a way, always streams out from the I-pole. Therefore each decision that I have freely performed from myself not only leaves behind a sediment of habituality, a persistent force of inclination and, if I follow it, a growing force and likewise a habit of deciding in such a way; but rather, as I, I am in my decisions and am enduring I as the subject of my abiding convictions and rational I as the subject of my free convictions: which I continually recognize as mine like I recognize myself as identical I in remaining consistent in the direction of my decisions, being true to myself in them and rational with a consistency that can be rationally legitimated at all times. The rightful talk of legitimation and self-responsibility applies everywhere.

But as much as all acts, all judging, valuing, practical acts, and corresponding convictions, decisions (judgments, values, deeds, resolutions) are I-centered and have their origin in and from the I, not all acts are I-centered in the same way. This already is a difference: in valuing, "one gives oneself" more or less perfectly to the values. But is this different from enjoying the ever deeper and richer performance of that valuing in which the self-giving of the value takes place progressively and in increasing perfection? This concerns values that have wide and indeed infinite horizons such that allow one to pour ever richer streams of evaluating feeling out of oneself, to become "enthusiastic," while following these horizons by opening oneself to them. The more that happens, the greater depths of the I itself enters in action. The I is a pole, but not an empty point. The I is not an empty and dead substrate of properties, but an I-center of actions that itself has its depths. This means that it is an I-center that grasps more deeply and achieves more deeply, gives ever more from out of itself, displays itself with ever greater sources and in this display gives itself to its object.

It is striking, however, that the I is not just a polar, centering interiority, and thereby an interiority that achieves sense and value and deeds from out of itself, but is also an individual I that has a deepest center in all of its objectivating, valuing in feeling, and deciding; the center of love in the pre-eminent personal sense. It is also an individual I that in this love follows a "call," a "calling" or "vocation," an innermost call that touches the deepest interiority, the innermost center of the I itself and becomes thereby determined for new decisions, new "self-responsibilities," self-legitimations. One may say: values that have already been anticipated or intuited are that which call the I. This is preceded by an "instinctive" or open, [359] disclosed enthusiasm which is directed at beheld values, still undisclosed or disclosed, and then at spiritual values. But enthusiasm is not everything. There is a call to the I, a call to a definite decision that is absolutely binding on the I. Enthusiasm awakens the deepest I, not only, as before, to decide and be rational in judging, valuing and acting, to pursue the true, the beautiful and the good, but rather to decide in a new manner for certain individual, objective values or, rather, goals; or a call to the I to decide for certain objective-ideal fields of value, infinities of values, to put itself in their service, to hear them as absolute oughts directed at the innermost I, and to live according to them infinitely in absolute habituality.

The I has individuality insofar as it has an innermost I to which callings go, callings that belong to it, callings for this I. It stands under absolute personal norms as this I. It is a free I, insofar as it can follow the call that has been heard or refuse to follow it or simply not listen to it (turning away from listening, looking away). It should be noted that this individual call always goes back to values, but can be instinctively blind like the originally blind motherly instinct. Also in this way, it has the form of an absolute call and is experienced as absolute ought.

But are not new considerations necessary here? Namely, do we not face the question about the <form> of a higher legitimation of the instinctive absolute ought through reason, insofar as reason discloses the value horizon of the instinctive ought and delimits its rational goal and clarifies this goal in its best sense? Yet another function of reason. Does each I have its absolute ought in the sense that each necessarily hears such calls, whether he follows them or not? If this were so, then the individual "ethical," "moral" demands or duties would be classified according to their typical directions, and universal norms would be recognizable for each I with regard to their individual duties. And if each I had peculiar individual duties that could not be compared, then it would be a universal law that everyone must follow their "conscience" in all circumstances.

Chapter 3
Husserilana 42, No. 27: <The absolute ought in the development of humanity and in the life of the individual. The new categorical imperative.>

Edmund Husserl

<§1. Rational willing and irrational (instinctive) willing. The irrational in the absolute ought. Rational grounding of the irrational.>

Rational willing: I want the better and the best of what is practically possible. If I choose the lesser good or remain passive, I am stupid.[1]

Irrational willing: I feel the instinctive urge to take care of my child's well-being, to give it every possible sensory pleasure as long as it is a growing child and its mind has not yet awakened. "More closely considered": "Because" it is then healthy and grows strong. The sight of its physical becoming is a "good" that I promote. The child's sensory delight is indicative of its flourishing. *But at first it is instinctive*, direct, irrational, evaluated and demanded without reason. The end of the development is disclosing experience and rational consideration. One soon goes hand in hand with the other: experience and instinctive demand. Sensory pleasure of the child that carries the consequences of pain with it, and also carries physical deterioration with it. Physical flourishing is instinctively enjoyable.

[1] Values of delight, principle of summation, absorption. Values of ought, ethical values—are these not the only true values?

This chapter was originally published as Text No. 27 in *Husserliana 42*, 383–391, dating from the 1920s.

E. Husserl
Freiburg im Breisgau, Germany

S. Heinämaa, A. J. Steinbock (eds.), *Values of Love and Ethical Reflection*, Husserlian Legacies: Themes for the 21st Century,
https://doi.org/10.1007/978-3-031-68698-6_3

Then "psychic" awakening. The child becomes a human being. It should become a beautiful <human being>. Instinctive liking of the human being; but at the same time, being lovingly directed to a becoming that increases in value. Each value that I have as a conviction and that can be projected onto this child's development becomes for me a *task of obligation*. In this I can, [384] of course, act very mistakenly insofar as perhaps my value-intentions are mistaken, my estimations of people, and so on. I value positively every human development that leads, in the most undisturbed and best possible manner, to a value intended by me. This pleases me. It also makes me glad to help. But it is different with my own child. He is "*entrusted*" *to me*; and I have duties here that I must not neglect.

I may also have duties to others. The other is in danger of going astray. I take precautions, I save him from sinking. I feel here: I should take action, I must not merely watch, and so on. Every original absolute ought entails something *irrational*. Does this not have to be sharply emphasized? But does not every human being become the theme of absolute ought for everyone? So fundamentally distinguished as value from every extra-personal value?

I cannot dedicate myself specifically to all of them; the practical impossibility separates us here. But of the people in my surroundings, my child is the "nearest" to me, and therein lies an irrationality of the absolute ought.[2] Other people occasionally enter my circle and possibly also become the nearest ones to me, provided that an absolute ought does not demand preferring any another as nearer and the nearest. Does the nearest determine simply the practical possibility? Or can the instinctive preference be grasped as an auxiliary moment that provides rationality? Love for science and "dedicating" one's life to science. The artist and art, and so on.

Does the irrational demand rational grounding? And what does that mean?

Each instinct is blindly directed toward a goal. Once the drive is disclosed, it sees the goal that would want to become awoken and would want to awake, and yet it does not see what it is "actually aiming at." This means: even after this disclosure there is something to be disclosed that perhaps the instinctual subject itself would never be able to understand (to disclose). The animal does not know anything about "conservation of the species." And "conservation of the species" is a title that perhaps "means" still more in the context of the world and demands further disclosures. [385] Admittedly, this is bold talk. The main point lies in this: if the subject of instinct were capable of disclosing the instinct gradually, step by step, then it would have to revert to chains of values and would have to want, at each stage, not merely the next disclosed goal, but the goal for the sake of the value, and want these values for the sake of still higher values, and so on. Is such endless disclosure possible with respect to all instincts? And <is it> not the other way around? The ought will be legitimated to the extent that the disclosure of the connections between the steps displays something rational. Even evidence of the ought emerges then as absolutely directed to me.

[2] Development of the absolute ought: the familial ought, the national, the humane. The Christian love of neighbor. But instincts of flight, instincts of revenge, jealousy, and so on?

Let us take the case of the human being. The instinct of parental love is disclosed for the human being's thoughts, but only up to a certain extent. And as far as the instinctual love is in their thoughts, they will rationally what they will, for the sake of the insighted value—whereby, however, the instinctive ought always remains with a non-insighted content. Likewise, with *acquired instincts* which grow as unconditional commitments to a vocational goal in the development of communalized people, out of unconditional love for science, art, and so on. Science is a title for value-formations, for example, mathematics. Love for mathematics is love for these values and the resoluteness to get to know them and to form them as inventor by generating them oneself. But these are common goods, and science reaches beyond me and my formations. I take pleasure in what I have formed, but I increase the "treasure of science." I perish, and science lives on. I see that the product I have generated has value not only for me and for those who may enjoy it at present, but that it is a value element in the abiding science and is called to enable higher values in science and to remain lifted up in it as a foundation that is neither expendable nor a particularity that can be extracted at any time as a deductively self-evident consequence of the higher universalities. Science is an open system of becoming. And for each present it holds that the system of science is simply such a basis that it is a condition and stage of an "endlessly" becoming system.

But science can be disclosed as a "necessary" function of humanity in its value-directed becoming that, in endless progress, [386] is at the same time an end in itself and a means for higher values. Mathematics for the other sciences, the sciences individually for the universal science.[3] This embraces the whole of human life in its value-formations and opens up prospects for imprinting higher values on human life and its environing world. Thus, science would receive a further *sense of disclosure*.

This concerns all "instincts," all blind directions of striving that appear with the character of an unconditional ought and willing. The disclosure is carried out together with evident establishment of norm that determines and delimits a rational sense and, on the other hand, advances from rational sense to new rational sense, from the fields of one ought to those of others. And finally, through the whole of humanity, in its development, in its history, runs a *unity of absolute drive*, the disclosure of which does not merely establish a fact, but makes the fact rational and comprehensible: this is a hidden rationality, though perhaps imperfectly self-legitimating and such that it generally guides humanity as if it were guided by a rational insight. It will so guide, however, after critique in the form of an absolute ought. In what does rationality consist? Not in the blind, general orientation toward

[3] But the useful sense of this discussion is only this: Every original and <every> acquired instinct is blind. But when value deliberations are introduced by reason and the immediate and mediate connections of values are pursued, the instinctive drive, which has a blind absoluteness of "I must," acquires the character of an insightful "I ought," and this is necessarily directed to values in the evidence of the very "I ought"—they are entrusted to me, and so on. It is thus said that the blind drives, original and acquired, are not entirely senseless, but contain within themselves, as it were, hidden or already conscious value-intentions, are subject as such to criticism, and through it pass over into an evident absolute ought with a personal sense of value.

a universal system of tasks to be determined as absolutely obliging, toward which humanity must live in order to be genuine, in order to be able to do enough for itself? [387]

<§2. Disclosure of the reason hidden in and governing the development of humanity and disclosure of the ideal goal of striving that guides this development.>

The essence of humanity is to perpetuate becoming as humanity by striving. And a unity of striving is at the same time unity of a *striving for unification in concordance*. Self-perpetuation that prevails through all disruptions and conflicts, obtainment of inner equilibrium, of satisfaction. Its essence must now include stages of development, forms of relative stability, and must be attuned to them.

But once the stages of humanity exist and those of scientific humanity, there arises from them the possibility of *scientific disclosure and self-understanding*, attuned to universality, that is, scientific consciousness of what humanity is as humanity, of what it strives for as humanity, what it instinctively would like to constitute, of what it could ideally bring to concordance with itself, to practical satisfaction. If humanity has risen to self-understanding, then it gains the *absolute norm* insightfully within this understanding, and so does every individual who co-understands.

Humanity is thus aware of the *reason that obscurely governs* it[4] and has now disclosed its own evident universal ought that insightfully brings all particular oughts to unity and to the unity of reason which is comprehended and fully understood as the last ought; this ought of reason does not, however, expose any absolute ought that each ought would serve and from which each would receive its normative sense.[5]

The disclosure of the sense, the "last sense" of the factical human life, is the disclosure of the sense of the world, and <this> is the absolute norm of life and of the world itself; it is their absolute truth whose being lies in the endlessness of realization. As absolute norm, it is absolute ought. Only the limited ought has irrationality; what is total and disclosed is rationality through and through. [388] We cannot do otherwise as human than to affirm in the will and first of all affirm universally what evidently turns out to be that which we actually aimed at all of life, what we had to strive for through all willing, no matter how sinful: a greatest possible concordance of fulfillment of all our striving, directed at the idea of a perfect unanimity; or toward harmony, inner unity, pure contentment, through all decline in sensuality and sensual delight, in error and sin, through all the concomitant inner brokenness,

---

[4] Reason as if, insofar, at least, as not every acquired instinct can be rationalized.

[5] Yes, it is certainly not demonstrated—if one could demonstrate it! That was, after all, the futile effort of all teleological proofs of God.

disunity, restlessness, wretchedness. But that comes about in the then necessary ascent from pure values to higher and higher values and ultimately in the conscious direction of life toward the infinite, toward the value of all values, which strips all other values of their relativity and of the counter-striving by all finitude and alone gives them their last sense, their last value meaning.

<§3. The absolute ought of each particular vocation and the universal vocation of being human.>

But does not an unclarity remain? Are not all values only values as target points of absolute ought? The truth of judgment is practically obliged—for the scientist. In itself, it is a possible practical good for everyone who strives for possessing the states of affairs themselves through judgments; then, in respect to the ought, it is positioned such that their absolute ought points at this direction. How far is truth something beautiful? The self-possession alone is not anything beautiful, just as not every experience as such is beautiful. Rather, to achieve truth by striving toward it and to rise to higher truth by continually striving is beautiful, and thereby vague thinking and erring is not beautiful. Whoever loves truth and strives for it and makes it their vocation is moved by this beauty, and this value seizes them in the absolute ought.

To occupy oneself with science and to enjoy it "as a distraction" is an *appropriation of values without a higher call*, just like an aesthete without a vocation is continually occupied with beautiful things. Whoever occupies himself without task, sometimes with art, sometimes with science, sometimes with charitable events and the like, appropriates values to himself but *not personal value*. [389] On the other hand, those who work can enjoy themselves, have an instinctive urge for psychic relief, in which they dwell in the land of beauty that enriches them. The urge "I must do something else for now, I want to and must recover" is itself an absolute ought when made insightful, founded in the absolute ought of the vocation. It is secondary and altering in as much as it does not habitually bind the ought to a direction. To elevate myself to great things, to be carried up above everyday life into a realm of the noble and sublime, is not enjoyment and belongs to a more general class of the absolute ought, alongside the fixed directedness of the habitual ought. Noble and sublime values include those of science and high art, not to mention religion, but also the values encountered in contemplation of great personalities by biographies, and so on. Once we have become aware of these values, the absolute call of the ought arises, although the size of perpetual life tasks may require restrictions. Then even the great and sublime must be "sacrificed" by me.

Do good! Do your best! But what does the "Do!" mean? One who sits in the theater does nothing, and one who "does" nothing else than things like this is rightfully called an idler by people. And yet, in his inner "activities," he perhaps creates according to others something beautiful and highly valuable, and in a certain manner appropriates it. Even one who is moved in this manner by the most noble and sublime is an idler. But one who is recovered by and seeks upliftment in beauty and finds it at the theater or concert is not called an idler, but rather someone who relaxes and educates themselves in time off work. *In addition to one's special vocation, everyone has the general vocation of being human*—if one is a true human being, whose truth is to want to be a true human being. Doing is everywhere *doing according to an absolute ought*, that is, in the universal of the ethical sentiment, which is the deed-sentiment, an ever-productive doing directed to the realization of the works that have been entrusted to and imposed on me, so achievements of all kind, and directed to the realization of myself precisely through these works, but also through their enrichments and elevations, which can be the sources of my true becoming and my true power in my specific vocation. [390]

<§4. Revision of Brentano's categorical imperative.>

Do the best among what is achievable! And yet I am not satisfied when I do so. Is this dissatisfaction unreasonable? No. I would not get rid of that dissatisfaction even if I had always done so and would always do so, and all others no less. It remains so as long as I and we all <remain bound> to the unrelated particular demands and their values. *The higher demand is in the doing, in the deed*: *Do your best, as that which is the best you can do in the sense of the absolute best at which the sense of your life too should aim, as is so for all people!* So, direct your gaze at the Absolute; it is what discloses to your absolute ought its ultimate meaning! Know that all value of ought, and the demand to help realize all absolute obligations of all human beings, has its highest, rational sense, the sense which makes it understandable, in a *world of God*! This is the now necessary rational belief in God.

Thus, Brentano's rule is insufficient. Everyone has their absolute ought. Their choice is carried out in the question "What ought I…?"; and where I ought to do several things the question is "Which is now necessary for me?" and not simply "Which, in the comparison of goods, is better?" Playing one of Mozart's sonatas is more beautiful than bathing the child, but the latter is obligatory because now is the time for it.

All practical goods do not stand on one level for me, not even all that I could realize. The voice of conscience, of the absolute ought, can demand something from me that I would by no means identify as the best in a comparison of values. What is foolish for the value-comparing *reason* is approved as ethical and can become the object of greatest veneration. On the other hand, every absolute ought itself entails its own source of highest values, even ethical values. To obey an absolute ought *simpliciter*, and precisely without questioning the resulting value or even delight, is an objective value and that which ultimately determines a value of the personality. Thereby it is an absolute ought for everyone, that is, a general law, one of an ought, *that everyone's absolute ought is included in the sphere of my absolute ought, in the sphere of my absolute ethical responsibilities and values*, [391] according to which

I cannot, without sinning, hinder anyone in acting for their absolute ought or even divert them from their duties. That is, such behavior is a particularly great ethical disvalue.[6]

Every personal subject is an ethical subject and as such has their universe of ethical values and disvalues. This ethical universe, however, is not a private matter. In the human community, all these universes are related to each other and form a single universal context. Also, on the basis of what was said above, each subject is related to this universal context, even though each has a differently formed universe, because what is immediate duty for this subject is not for the other, while the other is nevertheless ethically interested in this duty and its fulfillment.

And I already tried to show above that all absolute ought and the universe itself are capable of and in need of a "disclosure" and that this rational disclosure—in the form of rational consideration of human life in general—posits a highest absolute ought to consciousness.[7]

Husserliana 42, Appendix 36: <The unconditional "you should" that is not subject to a rational grounding. Departure from Brentano's ethics of the highest practical good.>[8]

Should the mother engage in such examinations of the highest practical good and first deliberate?!

This whole ethics of the highest practical good, as derived from Brentano and accepted by me in its essential traits, cannot be the last word. [392] It calls for essential delimitations! Vocation and interior call do not there receive their real due.

There is an *unconditional* "*You should and must*" that is addressed to the person and that is not governed by a rational grounding for the one who experiences such absolute affection and is not dependent on such grounding in its legitimate bindingness. This bindingness precedes all rational analysis, even where such an analysis is possible. "I would abandon myself if I acted differently," "I would never forgive myself"—in its case, this is what is at issue in all legitimation. The characterization "The weal and woe of the child is *my* weal and woe" is correct here, but further, the child's weal and woe is above mine. I can disregard my welfare and woe, but not that of my child. The welfare of the child is entrusted to me, I am responsible for it. To support its welfare is "absolutely demanded" of me. But the suffering of the child

[6] The concept of ethical disvalue, just as that value (value of an ought), would have to be especially elucidated. Of course, ethical disvalue is every doing against an absolute ought.

[7] Precisely the universal of revelation in individual orientation. <Cf. margin of previous page, namely, the first paragraph of [390] .> The ought can only have sense when it is the call from God [to me] wanting to will to co-realize his absolute perfection in the world according to my ability, in the *progressus* of value creation [*Erzeugung*]. Community in the absolute ought then becomes the greatest earthly duty.

[8] <*Husserliana 42*, 391–392.>

is still my suffering; namely, in its suffering, I suffer (based on my knowledge of it), and in such a way that all other suffering is secondary to this suffering. Eliminating suffering is not my goal, but the absolute goal, one and only, is to eliminate the suffering of the child, whereby my own suffering admittedly must cease. Is compassion an analogue of co-judging? Is it the same with co-valuing and co-willing? But the matter is not that simple.

Husserliana 42, Appendix 39: Two Ways <Universal reflection and the problem of suicide.>[9]

1) *Way from the praxis to universal self-reflection.* Starting point in preliminary reflections on human life, leading to the idea of a *novel universal humanity-reflection* which emerges from humanity's essential practical striving for self-preservation: the idea of a universal science. So, reflection on the universal structure of human life as an acting life, acting into a world, in the universal form of striving for "blessedness." And, thereby, the requirement of positivity: belief that finally satisfied existence is possible in the life which is directed at the environing world. Belief that a satisfactory human life can be carried out in the world, that "one 'could' live," that "the world would allow existence"—one can "exist" (*practical general thesis*). Conditions of possibility of such a life.

Way through the establishment of universal *positive science.*

But then further. Conditions of the possibility of a fully sufficient positive science in a purely theoretical interest—and that says: of a fully sufficient cognitive satisfaction (cognitive blessedness). In the science, therefore, the human being creates a methodological aid for the universal praxis that aims at finality. [450] The human being wants to carry out a particular universal praxis under the idea of "theoretical" finality in order to be at all able to lead a practical life of finality. This particular universal praxis, the scientific one, has a form and kind that "theoretically" encompasses all other ones and that can come to play in all others as means for their possibility. The simultaneously practical and doxic (theoretical) general thesis must be inhibited for the execution or production of a final and thereby responsible science. Exactly put: the universal theoretical general thesis already suspends the practical general thesis. This leads, first of all, to phenomenology.

2) *Another way of reflection starts closer to the fact of the given sciences themselves*, without any question of what they mean for human life in general. They are sciences of positivity, doxic positivity. The critique of the positive sciences, their essential restrictions and their naivety, their lack of "radicalism," leads again to the insight that a phenomenological reform is required, a new, completely universal and

[9] <*Husserliana 42*, 449–450; to Text No. 31: "Transcendental Reflection on the Whole of Human Existence and its Internal the Teleology. Development of Autonomy as Development of the Rational Idea of the Genuine Humanity.">

presuppositionless science. After that the conditions of the possibility of the positive sciences have to be researched more deeply. The grounding of the universal absolute science then also encompasses (in its task horizon) the universal theory which usually, in naive limitation, is called "ethics": that is the universal theory of the possibility of an all-human praxis in which humanity can lead a genuine existence. This ground also encompasses the whole of universal science as a structure of particular practical interest which then is called "theoretical-philosophical"; *and so philosophy (as theory) also encompasses—in reference back to itself—its own function for universal praxis*.

That is where the problem of suicide arises. Why should it not be conceivable that all human beings, including myself, in the conviction of the disvalue of existence, would commit suicide, kill the lower animals, destroy all other life beforehand, or that all other life would be destroyed beforehand and now the humans would exterminate themselves? Does my "death," which is only conceivable for me through others, presuppose the being-already-for-me of others who live? Does life not presuppose intersubjective unanimity of life, which carries in itself the naive and then conscious overcoming of all despair and carries in itself the development to genuineness as an apodictic necessity? Is that not the source of absolute teleology, whose necessary entelechy is God?

Chapter 4
Husserliana 42, Text No. 37: <Universal practical reflection on human life in its expectation of happiness and an enduring satisfaction of life.>

Edmund Husserl

<§1. The motives for the universal axiological-practical reflection on the whole of human life.>

The I, the person in the manifoldness of its periodically and aperiodically occurring needs, in types of needs that have become familiar. The lasting ends and means, the habitually available acquisition of goods and the ways to avoid typical inhibitions, evils, harms—is there a unity in all of this and <is the person> itself conscious of this unity, as it is apprehended as such by every other person in the experiential mode of empathy? But this experience of oneself and others in actual striving and living is closed to the horizon beyond the goal or means explicitly in view, beyond the realizing enjoyment or lack of it in the individual, halting only occasionally and reaching out and opening up in stretches of reflection. <An> exception is the expectation and continuation of a universal personal self-reflection, or also of a general reflection on human beings, how they foster living in the traditional commonality of a personal humanity, how typical ways of human being are to be distinguished, and how these influence the universal satisfaction of their life, their "happiness." But not as a "disinterested" thinker.

This chapter was originally published as Text No. 37 in *Husserliana 42*, 502–515 dating from the 1920s.

E. Husserl
Freiburg im Breisgau, Germany

S. Heinämaa, A. J. Steinbock (eds.), *Values of Love and Ethical Reflection*,
Husserlian Legacies: Themes for the 21st Century,
https://doi.org/10.1007/978-3-031-68698-6_4

Dissatisfaction in particular cases motivates the review of the reasons for the failure, and general dissatisfaction motivates a survey of life as a whole and, at the same time, its contrast with the happy life of others and the consideration of the reasons for their happiness. The universal context: the traditional unified national community in general need, which includes the following: [503] the unhappiness of the individual does not present itself as an accidental personal fate, in contrast to the normal well-being of the majority and of the whole people as a people. Motivation for universal reflection on human life and human welfare.

The human being, accustomed to and practiced in the struggle for existence in normal, narrow environing worldliness, surveys the world, surveys humanity, to which they themselves belong, in the interested attitude. In the survey, as in the restricted situation, the question is: "What is to be made of it?" This is so in the most general situation of the world and humanity.

But here something new arises. I am no longer the one <living> in the scattering of individual purposes and systems of purposes as they have arisen for my life up to this point, in order to satisfy whatever needs and purposes are most pressing among my ongoing ones. My entire future life, and with it also my past, becomes questionable. For, now, what becomes co-questionable is what I must hold onto of the ends and acquisitions of the past, rather than surrendering them as harmful, considered from whole life as a whole. In the universal axiological-practical reflection, I put myself into a halt by way of phantasy, as if I could "make" my happiness, as if I could willingly realize my happiness in the total situation, my lasting satisfaction of life; this is similar to how I, in the finite situation, position the will within the as if and ask about possible ways of realization. Therein lies the following: How must I orchestrate and order all my life in the possibility, thought of as my real possibility, that all my individual volitions and actions harmonize together into the unity of an undisturbed harmony of total self-preservation extending through life? How do I make these volitions and actions felt in the unthematic reaction of the heart, of the constantly accompanying mood of satisfaction, as an acquired, never to be given away, constantly available possession and in the hope to be able to expand it and to advance upwards in the future?

If we think of this reflection as successful in its expectation, the result would be the following: I, in a willing life, going forth into the infinite and whole, would be governed in all individual volitions by a single habitual will, precisely the one which, as the supreme regent, would have to give each singular manner of comportment that form of happiness. [504] Although the whole consideration proceeds imaginatively, all the singular and systematically connected possibilities that are considered are of the valid form of "actually, I can, I could." Also, if we posit the case of my actually coming to an intuitive form of the result, that is, to the concrete evidence of "I can do so," from then, the "I will" would then necessarily arise from it. Thus, I have become a new kind of I (or would-be I), the one that has shaped itself out of the form of the familiar and relative human rationality to the realized pure rational-I to that which, having realized itself as rational-I out of an archontic will, is complete. –All this is still hypothetical. We do not know whether it can indeed lead to such a result as evidenced possibility and actuality.

<§2. The question of happiness as a question of an axio-practical life affirmable as a whole.>

Now, however, it has to be considered that I am what I am, and <that I>, in my personal <life>, that is, my life of my needs and purposes, live as the one I am, not in isolation, but as being in such a way, and only able to be in such a way, in the social and, ultimately, in the human bond of community. My life is life in my We; and our life is thoroughly a living together, for each other, against each other; which is to say, it is not a life in external collection of being together in space and in time, but rather a life of interior participation, in experience, in thinking, in caring, in love or aversion, and so on, in the life of others. Precisely because of this, the weal and woe of each is dependent on that of one's fellow human beings—as it is in the particular situation, so it is in the whole of humanity and its history.

How, I ask, must I shape my (whole, broader) life in order to gain a constant life of satisfaction? Of course, an ideal of satisfaction that is already fixed in advance is not to be taken as a basis—as if I had in mind a wishless life, a life of satiety, a life of constant delight in which desire is fulfilled constantly and without further ado, or as if I had in mind the constant success of all particular aims and the like, [505] as though such a life, under the idea of a completely open, endless, and constantly proceeding life in such form could result in total satisfaction. It is one thing that the fulfillment of individual wishes, either those that I have from earlier or those that are just being stimulated, yields satisfaction in the particular case; it is quite another that the general mood of a total satisfaction, with which there is given certainty that an enduring life proceeding in this manner, is evidently desirable and which I, reflecting on it, could and would have to affirm practically. It can be that I would have to recognize that the earlier fulfillments, delights, supposed values of happiness would have to reveal themselves *ex post*, for later life, as unhappiness, and so on. How can a human life proceed in such a way that it would be axiologically-practically affirmable as a total one, in which there would be no need for remorse, for reversion, for the relinquishing of the supposed and acquired good, or, if that were the case, then only in the manner of correction, of ascending improvement? But is the question not already answered, and in the sense of a complete impossibility? The question of happiness as a question of the power to make one's happiness through reason leads to a universal world-reflection. First, as a reflection on how the world as such is, being the world in which we humans live and in which we are to shape our wishes for happiness, in accordance with our actions. Those of us with this wish, if we currently have it, belong to the actual world, and, alongside us, other human beings in all the ways in which they thoughtlessly live out the day. We, as subjects of the world to be shaped, have other people opposite us who, while they are such subjects for themselves, are objects for us—means and material to be, perhaps, shaped according to our interests of happiness.

<§3. The question of the possibility of a happy life and its conditions.>

Does the world—and the people in it with their original and acquired capacities, with traditional and anti-traditional ways of behaving, with their passions, follies, and so on—fulfill the conditions of possibility for any human to shape their whole life according to reason? [506] To what extent must the world be investigated as a world of matter of facts, the thing-world, of the near and the distant, of the world of the animal and the human according to their driving and doing, but also according to their capacities of intervening in the course of the world?

But the question is also: the talk of a satisfying and dissatisfying, of a happy and unhappy life is initially vague. It can have its experiential right only if we are already guided by examples of experience in which we can recognize or construct the actuality or—by idealization—the evident possibility of a happy life, of an envied and enviable life (one that, considered as a totality, is happy). We may be mistaken in fact, but we may, nevertheless, have images of an evident possibility of such a life, in evident construction. Of what kind are they? We know that there is misfortune that threatens everyone everywhere; we know that it has many forms, among them those that can neither be foreseen nor avoided by reason. Nevertheless, a life can be enviable and, despite all the accidents in the particular cases, despite difficult trials, still be happy as a whole. Sustained successes, acquisitions of the past, constant hope in the future—this is a "matter of luck"; fate has, in fact, in individual cases and in certain narrow circumstances, even foreseeable ones, let it happen this way, without any regulation of life by a uniformly governing reason, let alone by something like a life-plan and the consistent execution of an identical monarchical will. On the other hand, experience seems to teach, within the limits of finitude, that the cultivation of reason, the reasonable knowledge of the situation, and the well-weighed probabilities that are prospectively calculable, make it possible to avoid serious misfortune that would otherwise rule out happiness in life altogether. To the traditionality of human life, in which I find myself and we always find ourselves as mature human beings, there already belongs that, in stretches of life, and extending through whole vocational stages of life, a life of success uniformly centered and ordered by practical reason is possible; in its steadfastness, a life of resolute hope and thus of "satisfaction," a life to which one says, "yes" in one's will, in explicit retrospect, is possible. [507] The thesis of the will proves itself in the evidence of the anticipation of capability, which evidently implies future capability.

But the nature of this evidence must be considered. Inhibitions, disturbances in the form of unforeseen and unforeseeable factual contingencies or personal fate can very well occur: illness that weakens me, that makes me unconscious, that paralyzes my memory, that does not allow me to know again, that does not allow me to will again; on top of that, there is failure of the fellow humans with whom I am intertwined in the willing life, not to mention the fact that, in their existence and behavior, they also, outside of all social connectedness, determine my environing world on whose normal course my foresight is built. My acting life is always in an environmental horizon, as is my vocational life, which is related to the wholeness of life. It is the life of the will in its horizon of the will. And this is a horizon of beings that, in each case, is familiar to me, namely, of that which, as I know, is and has its relevance for my acting, and a horizon of the goals toward which I am directed, in me and through other co-participants (servants, co-workers, and the like).

Traditionally, human life is stratified, in a normal manner, into vocational unities, into unities from a centralizing will, into spheres of duties. In each, there is the acquisition of goods, horizons of hope, but they all interrelate in such a way that even unhappiness in one can reduce the value of normal happiness in the other. I therefore count physical welfare as a special field of first instinctive, then volitional care that extends throughout life. Everything else depends on health, which requires a harmony of needs of sensibility to be regulated by will, which appropriately restricts and arranges them together. Furthermore, one has social instincts, social (typified as traditional) goods and duties—one has one's sphere of family duties, one's sphere of family life as a sphere of regulation by the will. As a citizen, one is obliged in service of the people of the state; as a civil servant, one is obliged in a special way as an organ of the government. The concrete life of the human person can be "happy" only if, [508] with regard to all these layers of life as particular layers of the life of the will which also mutually bring each other to the fore and condition one another, a certain style of harmonization together into a whole of life is ensured.

How does this look in the traditionality of a normal historical life? One lives in community <with others> and lives with—giving oneself over more or less intimately, though possibly also fleetingly and quickly glancing away—their misfortunes, their blows of fate, and sees how they behave.

Death is an unhappiness of which no one can be spared. I do not mean one's own death, but the death of the beloved. And everyone (in a normal manner) has people whom one loves as close relatives or as friends, as revered exemplars, and the like. Losses of loved ones can deeply shake the will to live. Some are broken by the loss; they live on, but in despair, in hopelessness, paralyzed in their life activity. For others, the loss is the most serious, dangerous wound that time heals; they take the fate upon themselves, overcome it and later live a healthy life again, one that, for their own aspect, is successful on the whole and affirmed as worth living. In certain cases, as in those of the most intimate and deepest inclination for love, death has the character of the absolute annihilation of the whole value of life; the lover has merged their whole being and life in that of the other, has given themselves entirely, and, with the other, their own life is annihilated. And yet, is this not an exaggeration? Does not everyone have his own, original duties that cannot be abandoned, duties of loving care for others that, for the most intimately loved one are only secondary, sympathetic cares and cannot have the character of original ownmost duties? As the lover naturally concentrates their activity on the loss, it becomes the activity of suffering. The sympathetic third-person is shaken, but already out of instinctive self-preservation, turns their gaze to the wider context of life and value, in memory of others who have suffered, who have overcome and withstood. [509]

<§4. Love and sympathy. Enmity.>

Love in the excellent sense is a personal being-with-one-another and living-with-one-another with the horizon of the whole life. It is not merely being and living as an individual-I and only occasionally living together with another, but rather being in oneself, at the same time, in the other and bearing the other in oneself, and this in a completely concrete manner, that is, insofar as the I is just a willing-I, it is a volitional being in the other and his willing according to all modes of willing, such as are in all acquisitions, possessions, habitualities. The other is constantly in my life horizon, as I am constantly for myself, and in everything that may affect and occupy me at any given time. The other is always "present" for me, like how my past is always "present" for me, even if I do not explicitly remember it. Although the other is always inseparably united with me and "present," he is only so in the manner of empathy and empathy-modification. And my being-one with the other and living with them in volitional devotion is horizonal; I only participate in it through possibilities of communication with the other, but my universal life interest is at one universally for the life of the other as interior, egoic participation. My life-will has the open, infinite horizon of my total future and, as at one with it, of the beloved's future.

So, this is an excellent case of a union of love as a concrete identification of one's own being with that of the other and, in mutual love, actual and reciprocal identification. Much would still have to be said about discrepancies, where the other "forgets themselves" and does something with which I "cannot go along," and so on…

Each volitional living together in and with others is love and is, so to speak, a segment of the I-You unification, of identification with egoic, so volitional being. In particular, there is concurrent "harmonizing" with the others in the connection of empathy; it also constitutes a concept of sympathy (here as actual). If, as a rule, on the occasions in which we come together with others, we are in harmony with them, and if we have the foresight, the pre-consciousness, [510] that we would harmonize with them at all (possibly immediately the first time we meet, having an instinctive fore-feeling), they are, it seems to us, "congenial." People, however, can appear uncongenial to us, repugnant; empathetically, we understand them, but we do not interiorly go along with them, do not "take part" in their interests, in their joys, in their sufferings, and so on. This can go so far that we get the opposite case of the full loving inclination; instead of friendship, we get enmity, the total negation of the being of the other in all their life goals, in all its activities. Negation is thereby an intentional modification of positing. Normal human existence as existence in community is community in harmonious concordance, but community life occasionally takes the turn into discord, into occasional and also enduring antagonism and enmity.

Loss of a possession. The habitual direction of will towards the possession assumes the mode of regret over the non-existence of that which is still desired. Devaluation of a possession. The attitude of my will towards the possession as positing. My estimation of it as a good has changed into something crossed-out. Perhaps

I doubt. Here I am at odds with myself, my present self says no to what the past one values.

Conflict with others. First, the sphere of the action that transforms the real environing world brings conflict; what one transforms, the other cannot. Likewise with the consumption goods, food, and so on. The satisfaction of the instincts of the one comes into conflict with that of the other. In the inclination of love: I strive for something, attain it, and have the joy of attainment. To be absorbed lovingly in the life of the other—in their attainment, their joy in delight—is to be delighted oneself in the other. But even in the uttermost two-in-oneness, the tension between the mutually incompatible acts of I and other I. I am, as original I, at the same time, submerged in the other, in the other's volitional I, such that the one will and the will of the other are in conflict. They cancel each other out; the action is impossible. I can give up my willing-doing. As the other acts, I also come to act together in the other. Or the other gives in; I take note of it, I am originally acting and the other is co-acting in participation. We negotiate, or possibly decide from the common situation without further ado, without, for example, speaking further, [511] saying that it is "more advantageous" for both, if the other does this matter and I take on another one that is farther from their interest and ability or which concerns the other only on my account, or vice versa.

If they are not friends, two bound as one for life, this can result, as so often happens, in action against and to the disadvantage of each other. The one does something that makes it impossible for the other to attain an intention, but makes it possible to put hindrances in the way or, in the most direct manner, prevents their partner from acting through physical coercion. Instead, also threatening, communication of the will to do the other harm, to disadvantage them, and so on, if he carried out an action.

In general, one does not know the situation of the other, the overall state of their life-interests. In general, one does not seriously get involved in the life of the other and what must be important to them, what is for them a fixed direction of will, what their interest is that gives their being its state and core. One understands the other only in an empty manner, as someone who wants something that one wanted oneself, and who, on account of this, feels inhibited; in general, one is pained and unwilling to endure the disturbance by one's companions. But if one enters into the other and their situation of interest, if one thus understands them more concretely, one may readily see how important for them the action in question would be and, on the other hand, how much less meaningful for oneself. Thus, sympathizing with the other in empathy, one may gladly and of one's own accord inhibit one's own action, that is, be friendly, loving, to do something pleasing for the other, or possibly arrange, in the sense of friendly understanding, an action that is alongside and partly with one another that proceeds harmoniously. Every abolition of a conflict, every elimination of unpleasant struggle, of unenjoyable sentiments of others is beneficial. The inhibition of a will is painful and inhibition as an alien will directed against one's own will arouses a hostile mood against the other. Every understanding of the other as already hostile arouses in us a negative reaction of feelings against them. To live in disharmony with others is to live unhappily. But can it not also be a delight to be a ruler—a ruler who conquers all, bending their wills under his, purely by force, by power that imposes itself and its will at all times? [512] Power that makes struggle among the subjected impossible, and in general enforces a common life in which no one interferes with the actions of the other. A people under a ruler and a rule of law established by him through power that (as a rule) makes a practical common life possible without disturbance. Order of will by power and laws of power (law). Another order of willing: that of the harmony of willing.

<§5. Communities of love as communities of being for one another. Altruism and egoism.>

The human being in the community as "egoist."

In the community-totality of communities of love—be it in a serious or playful life. The community of love in playful deeds, in the common play but not love-as-if.

The family—the unity of a generation of the living (unity of a kinship), ancestors, the last living grandparents and great aunts, and so on, unto to the youngest, children (siblings, cousins, and relatives).

The normal family as community of love, unity of "harmonious" living together: everyone loving everyone, pleasing and wanting to please, "to prove pleasing" to the other, being pleased in their being, witnessing their deeds and letting be, but also making the other's matter their own matter, even engaging it, helping with it, and so on. The family instinct or drive, the "amusement" in harmony, in the activity of family love as fulfillment of this instinct. Primal form of love of neighbor. Breaking of the harmony in conflict.

Inhibition in the effect of a wish, an action possibly already undertaken is dissatisfactory, just as, conversely, fulfillment as attainment in uninhibited action is pleasant. Original inhibition of an action that is aimed at obtaining something, acquiring something, and having what someone else wants to obtain and have. They interfere with each other and they become each other's opponent—this, then, arouses negative feelings. To get along lovingly, to agree on deeds that do not disturb each other and do enough for both, to satisfy both at the same time in their individual and common situations, [513] these yield satisfaction for both and, at the same time, as an elimination of antagonism and dis-joy, yield a higher joy. From unloving comes loving harmony ("altruistic"). Egoistic and altruistic behavior: in the latter, one absorbs the other's being and willing into one's own, and, in the case of the incompatibility of actions, one renounces one's own for the sake of pleasing the other; putting oneself in the other's position, one prefers identifying oneself with him, their action, affirming it, co-acting "in" them; or, one decides for one's own action, but does so just to please the other, because the other prefers that. The egoist wants their own satisfaction, and even if they want to satisfy the other, to do something pleasing for them, it is only to enforce their own original will, to promote their own interests.

One's own interests—these can also include the interest in showing love to others whom one loves, this interest understood as one's own interest. Love lives in the other and has the final motive in the other's ends, in the other's welfare. (Loving care can thus undergo a reversal in which it is transformed into selfish care for oneself. Thus, what is love in the family and actually has the character of ultimately doing <something> for the sake of the other can take on the character, possibly peaceful, of familial egoism, in dealings with strangers and families, namely, through the explicit awareness: I want to be loving to my own, but not to strangers, who do not concern me.)

Everyone has a circle of one's own. In the community as a community of love, one's own is not given away at all, but each circle of one's own is brought into harmony with each other's own—into the harmony of love. Whomever has "one's loved ones," then also neighboring people for whom there is no love done, is, relative to them, an egoist and, relative to one's loved ones, a loving "altruist." But no one is actually through and through an egoist; at least, this is not normal, but rather that, for one's neighbors, for strangers that one meets, wherever there is no strong self-interest in question, one quite gladly does a favor, helps, and so on. Even if their own interest comes first, union with others does one good and one is happy; although one wanted to fulfill one's own interest and has fulfilled it, one is, at the same time, putting oneself in the other's place, happy that the other has found their direction. [514] This joy in the satisfaction of the other can really terminate in this and thus represent a component of love.

Evil egoism, malicious egoism, unbridled desire, and promoting one's greed at the expense of others, knowing full well that one is doing the other a harm, fully inconsiderate—this is another thing. To come to an understanding with the other—in malice, in egoistic lack of respect—then means that one does not go further than permitted, for instance, through fear of the consequences, in consideration of the laws and the power of the state, and so on, and seeks to gain the consent of the other for the sake of one's own advantage and not at all in the concomitant joy of doing them a favor at the same time.

Malice, wickedness, unrestrained egoism—are these, thought of as pure cases, not *limes*? Whomever is mainly or exclusively focused on themselves and their own sensible or spiritual enjoyment, do they seek and find satisfaction exclusively in it?

Let us think about the egoist living in the community. He has his "absolute values," final values; thus, he is satisfied in the attainment and knows no other <goal> than to order his life, to direct it in such a way that posits and realizes values of this type, which is absolute for him, as goals, keeping them in their order of rank. Will he really manage to reach his "blessedness"? He may say, "It is all the same to me; if I only achieve delight once, that is enough for me to give value to my life. Everything else is boring, indifferent to me; it may have value for others, but not for me. That is who I am, and as the one I am, that is <my> absolute decision."

The one who has an absolute direction, a type of goals that he decided for as absolute, reflecting on them, pronouncing them in absolute practical judgments, as judgments of his absolute ought.

Someone can stubbornly insist on an "error" "as absolutely evident for him." We call it "error," he calls it "absolute truth." Religious people of various religions and each in absolute faith. How far can mutual understanding and making understood, reconciling with one another, reach? [515]

I, the one philosophizing. And my understanding of the world, my understanding of human beings. My science, its "objective" truth: its "unconditional, universal" validity for "everyone."

Last, pure, absolute self-reflection, as ego in the world's validity, in the validity of others as having validity in me, as co-bearer of the world's validity that is valid for me, and so on.

Husserliana 42, Appendix 44: Ethico-practical reflection on the world and cosmological reflection on the world. <Cosmology as the whole of the factual sciences in the service of a truly satisfying life. Vocational attitude and attitude to the whole of life.>[1]

The human being among human beings, the human being in the generative context. Personal family, tribe, a people, a community of peoples.

The human being in love and hate, in friendship and enmity. In love: community of harmony, of inner unity and pure satisfaction in inner unity. Contingent inhibitions in human existence with one another and, on the other hand, hostile, egotistical inhibitions. Unhappiness in interior, personal disharmony, joy in the conscious production of harmony as the institution of a (deliberate) personal loving community, a unity of trust. In the individual case, in individual communities, an enduring community of love. Different forms of friendship and antagonism: individual, familial, standing with and standing against, people against people.

Human beings in the world of matters. The world of matters as a common one for humans as where the goals of the action lie, in which persons interfere with or support one another through trusting collaboration. Humans materially. In the world, humans over whom one can quarrel with certain humans, with whom one can compete, in relation to whom one can become suspect, and so on. The traditional environing world, taken universally as a world without this harmony, with necessarily individual instances of harmony. World of cares, possible disturbances, contingencies, hostilities, struggles and so on.

Eudaimonia. Self-reflection and world-reflection. [516]

Religion as a specific tradition. Traditional powers of trust, of help that make the cares of the world bearable.

People's religion—gods of the people, world religion—divine world powers. Development to the world religion—development of a universal cosmological view

[1] <*Hussserliana 42*, 515–519.>

of the world. *Universal anthropological*, axiological-practical *world-reflection* of humanity understood as personal *in contrast to the cosmological world-reflection* (non-axiological "merely being"). The human being as a person, as member of humanity as a personal universe. In personal life, the life of interests or the life of needs in the broadest sense. Normal: people, maintaining their "existence" by practically aiming at the fulfillment of needs and with the successful result that on the whole the person is satisfied in life. On the whole: that means that the human being is always hovering between dissatisfaction and satisfaction, between failure and success in their endeavors, between hope and fear, and so on. Overall mood of satisfaction as the normal one—in the explicating overview of life, affirmation, for the future of life, hope. Growing of the need for rational reflection and from it of a rational regulation of the whole life and the environing world of life. Normal foresightedness. Within the same practical possibilities, selection of the best, possibly the most probable—on the other hand, the open possibility of anomalous contingencies, fates, disappointments. The world, the universe of the *de facto* "being," is the real and possible material for praxis, in the sense of the sought-after universal life satisfaction, for human beings in their constant neediness and as those who undertake universal, practical reflection.

The individual human being is nothing isolated with regard to his willing-life, in their striving, in their needfulness; the individual human being is a member of the community, intertwined in humanity. Universal practical reflection encompasses humanity as a community, immediately and mediately, of persons who all strive for satisfaction and must come to a concordance, to overcome personal mutual inhibitions and hostility.

Opposite the axiological-practical self-reflection and humanity-reflection, community-reflection (that of practical reason), the *universal cosmology*. The human being, not as a concrete person in the living and striving for self-preservation, the formation of a living that could be affirmed, but rather in the cosmos as merely "being." Abstraction from all axiological consideration of humans and the world. The existing world contains humans among the animals, <among> the heavenly, the demonic powers, under things and their forces, with their own normal and anomalous drives. [517] The existing world contains human beings as they are factually, individually and with one another, as mere facts—exterior value, as rational and irrational, as fools, as sages, and so on. So, cosmology as the <entirety> of positive, factual sciences, including those that study human "matter of facts," that is, human needs, human endeavors to satisfy their needs, human reflection on the best possible life, on obtaining their blessedness, human endeavors to reform humanity as a whole on the basis of universal reflection on human life and its practically possible influence: all of these *as matters of facts of the existing world*. The attitude of cosmology, or of the human being who pursues it, includes the needful attitude of knowledge, the attitude toward the matter of factual, toward being and being-so and modalities of being, as mediation of the possibilities of my and our weal and woe, without questioning what I value. More exactly: the scientist is in a vocational attitude, a special "infinite" one, the attitude that encompasses the endlessness or wholeness of their life and is thereby in "infinite" living community with their vocational companions. This is a *specific practical* (and in this sense "infinite") attitude. There are many "vocations" in communal life, and the same person can have several of them.

Beyond that, however, every human being has other needs, whether regularly recurring or random. Everything and everyone contributes to the mood, the general satisfaction or dissatisfaction of life at any given time; it is part of the concreteness of personal life that it has unity in the manifold of needs and the activities that correspond to them. Individual reflections on that which already is in the respective praxis (or a praxis to be put into play) as a matter that can be practically formed, generally belongs to the praxis itself as activities serving its interests. Universal and vocational striving for true being and predictive truth is its own universal practice as a particularity in the concrete life and striving of the theorist, one that necessarily goes beyond it.

In vocational life, the vocational person splits, so to speak, in the person. I as a vocational-I in my own unity of life, in activities and habitualities in the concrete whole of life, of a vocational life that as such gains its own unity of universal *satisfaction* or *dissatisfaction*, its own happy blessedness and unblessedness, its own system of ability, its own system of ends, its own system of acquisition. But the human being is concrete, and this possible blessedness is only given to them in their vocational attitude; [518] in their whole life and in its entire state of happiness it is a relative and partial thing that contributes, that can outweigh and can be outweighed by other moments.

The choice of a vocation, as related to the infinity of life, has the peculiarity that it is a willing that encompasses an infinity of willings. As such, such a choice of a vocation encompasses actions of life in general, in the manner of a general will, as a general will to those vocational activities that are imagined in advance in an indeterminate-general manner, according to types of order and form.

In contrast to the various vocations and their choice, which is instituting for the vocational person, *there is a decision of the will, a choice that does not retain a special life as a special infinity in the concretion of full life*, but rather *the full and whole life*—as a will encompassing this whole to shape it "happily." Just as the originally instituting vocational will is preceded or can be preceded by a reflection on the "value" of such vocational life for me (whether it has the nerve of the ought of the need appearing in unconditionality that must be satisfied or, at least, whether it has an "attraction," predominant in contrast to the other possibilities of life, and, as such, deserves and demands practical preference—but also whether my strength would be sufficient, and so on), the "ethical" decision to the "virtuous" life; or, to the reasonable one, directed to the highest purpose of satisfaction, the decision can be preceded by a corresponding reflection on how it stands with the practical possibility and then the practical necessity, and so on.

Within the concretion of the I, as living in a multitude of habitual interests that have already been established, an infinite interest can come to be established that *generally has the theme of the regulation of all life and all interests as its subject*, and in its service, again, a general interest that has the sense and possibility of such a regulation as its theme. If a universal *theoretical* interest has already been instituted as cosmology, this is now in service of the unified and all-encompassing interest of "blessedness" or a universal will of life, will of self-preservation, directed at universal concordance while overcoming all existing discordance.

But now we do not have independent vocational research as cosmology, but rather *cosmology functioning for self-preservation*, functioning for a further vocationally established universal reflection of the concrete personality directed towards "true self-preservation." That indicates, however: we have, first of all, the *science "cosmology,"* [519] the science of the existing world and the human beings actually existing in it; secondly, as superordinate to it, we have a different *science of the essence, of the possibility, of the conditions of the practical enabling of a "true"* human *self-preservation* or a life to be regulated volitionally with the aim of permanent satisfaction. In contrast to the theoretical doctrine of the factually existing world with factual humans, their factual goals (among which are those humans who factually set, reasonably or unreasonably, themselves a life goal of satisfied life) is a *theoretical doctrine of possible human beings and a possible humanity*, which set their *true self-preservation*, a truly satisfying life as such, as the *ultimate goals*, with the sense that the factual existence of such humans or humanity is not in question. It not does not make a difference whether humans have thought of something like this, have put their whole life with all its special purposes under such a final end. Rather, in question is the being of the possibility of a human existence, in which the human being sets for itself this final end and becomes evidently certain of the practical achievability, and then also becomes certain of the need to seize it and to want to organize its whole life under it. *Instead of the being of the world, it is now primarily about the being of the achievability and of the necessity of the choice of ends.* Belonging to the factual human being in the world is that ends are set for itself, sometimes achieving them and sometimes failing to achieve them. In any case, therefore, achievable ends belong to the human being, shown in fact by the factual reaching for them. And, in advance, everyone has a horizon of reasonable factual attainment of ends (but also non-attainment, both mixed, but one and the other in particular inductively prefigured).

However, is the question of the real being of a practical possibility, of the goal and the ways to the goal, the question posed from the personal I as a reflection on one's own practical capacities and on the value of the goal for me, the personal I, a different matter? Instead of the question posed from the outside, whether, based on previous experience of one's own behavior, one is likely to have the capacity for the goal and the like, the question is personal: is it something worthwhile for me to wish for? Would I like to want and to attain that? And can I? Is it in the consciousness of my capacities, do I have, in the environing world familiar to me, practically possible ways to the goal?

And finally: the question of the unconditional value (the absolutely ought) of a goal that appears consciously as practically possible in the given situation with the character of the absolutely ought (and indeed as an *unum necessarium*). Here is an important problem of the will.

Husserliana 42, Appendix 45: Communal life and "existence." Life in the fatefulness of natural threats. Life under threat from other people and animals. Angst in existence, total existence.[2]

What is required is not only a single end, not only a habitual interest that, even as "life interest," spans the open horizon of the whole individual life. The merchant who is in danger of going bankrupt also says, "My existence" is threatened, that is, my self-preservation is threatened, I am going under, I am losing my existence, I can no longer preserve myself. If this really happens, however, there may be possibilities open to him to ground a "new existence," to take up a new "vocation" and to preserve himself in it; in other words, a new way to come to self-preservation. He has the possibility. A distinction must be made here whether others mean and say this or whether he himself says it, whether he "knows" himself in this possibility, whether he is aware of it as such, whether it is valid. If it is the case that he is aware of it, he still has "hope for life" and is in the care of life, in whatever form he "lives," that is, to preserve himself; however he can establish a new existence and keep it going, in whatever new form formed by him, by his willpower, into a familiar form of practical vocational life. In such a situation, he still has an open horizon of possibilities, namely, of capabilities, more or less definite and indefinite capabilities of "manners of existence." Among them in the horizon of possibilities, there are also open indeterminacies of the completely empty kind. This means: "I could (among the known vocational possibilities that are typical of the familiar human environing world) take up one or the other (typical, coming to mind) and maybe also others that are still unknown to me, whose practical accessibility for me I cannot yet survey." In general, the supposed capability is also indeterminate in that the ability to measure up to it; the capacity may not be fully secured but only be presumed. All life in hope is life in care of existence; and, vice versa, if care of existence is concern about the How of existence and is not about the That of existence as such. Only now, we have the difference that has just arose: the human being who lives in *normal* life, in the normal hope of life in general, [521] but, within it, is naturally in constant care about the How of fulfillment, about the How of ongoing life-preservation, has particular, well-grounded manners of existence, and these particular ones can be supported or broken. Nevertheless, there still remains a hope for life, unbroken as the will to live and as the life horizon of vocational possibilities, possibilities in a typicality of actions related to the typicality of the proceeding environing world and its mixture of regularity and contingency (fate)—approximately regulated, but in such a way that one can adjust and change accordingly. The opposite case is the rupture of the hope of life in general and thus the negation of all possible cares, the impossibility of caring at all. This lack of care is complete hopelessness; it is the negativity of life, which in its positivity is in hope and care, a life of directed, not isolated

---

[2] <*Husserliana 41*, 520–523.>

<care,> but a life in a universal and cohering horizon of capability, directed towards concordance of connection of successful active acquisitions. Alongside success there is failure; and alongside supposed success there is actual failure that is still in the process of revealing itself. Care is the mode of feeling that emerges from the modalization of activity and from the constant pre-figuring of the horizon of possible failure within which run the lines of still prospective and certain success, success under correction.

If human life in positivity, which gives a concise sense to the talk of life and the abilities of living ("How is it going?"—"It's livable!"), is thus to be distinguished from life in negativity, further questions arise. First of all, there are differences to consider, even for the normal life as life in positivity.

How is it, for example, with the life of a beggar? That is, one might say, an inhuman life, unworthy of human beings. The latter expression is a value assessment; the former can, however, express what the valuation is based on, namely, a human way of living that is, to put it more precisely, not human. One can also take the life of a prisoner and call that an inhuman life.

On the other hand, one will say: one should not look at the beggar externally and judged on the fact that he begs, that he feeds himself by begging instead of by his own work. The latter may simply be impossible for him, as in our time of unemployment. Seen interiorly, there is an enormous difference between whether he lives in hopelessness of being able to live through his own work, of being able to build up a continuously promising life from his own work or whether he has lost the will directed to such a life, though he does not hold it as impossible, whether he let his willing personality fall into decay, [522] whether, being weak willed, he has abandoned himself to temptations that knocked him off of his "path," out of the order of his willing life, or whether he, after some failures, again being weak willed, lost the desire to continue to make serious effort and to care. He becomes a "parasite," he turns begging into a vocation; this is also a "vocation," also a manner of organized self-preservation that enables existence for a prospective lifetime. And yet: Is it a true vocation? Is it a true, a truly human life? One may say so. Why this assessment as negativity of "real" human life?

Likewise for the prisoner. He is still in a human positivity if he can be in a "life" hope, in the hope of building up a uniform, active life, a life of active self-preservation, of active synthesis of his own achievement, in which he not only eats and drinks, not only satisfies "physical" needs, but, in the unity of a working life that is of a human kind, satisfies himself as a full human being. What every human being aims at as a "real" human must, of course, first be fully clarified, while here it can only be suggested. Now, what about if the prisoner knows that he is locked up for life and cannot muster any hope, seeing no possibilities for coming back again to such a life? If he cannot do anything "as punishment" in an isolation cell or if he is forced to do something senseless and purposeless for him, plucking wool and the like, without any serious possibility of escaping? Is such a prisoner's life not inhuman, not de-humanizing from the start? Would it not be true heroism to be and to remain human by maintaining the will to live with the horizon of a normal human life, a life to be built up with new energy after the end of imprisonment? From the point of view of atonement and improvement, does the penal law not impose a generally impossible heroism of the will on the person whose will is sick and weak from the outset? But apart from that, this is an anomalous alteration of normal life that the one who has blundered into transforms into a new higher normality or better positivity. For other human beings in a community who are ready to play around with their fellow humans in such evil ways, prison is one of the forms of fate, one of the limit situations into which everyone—whether guilty or not—can blunder into, as those that put the human in the greatest danger of losing their humanity, of not being able to preserve themselves in humanity. [523]

If one in such a situation invents games to occupy themselves and volitionally finds themselves "preserving" themselves through games, and ever again through games, would that be a human life? Is that not, so to speak, a manner of begging for life satisfaction rather than coming to a truly human way of existing? The life of the beggar in the usual as well as in this analogous sense is not genuinely human. Is it degraded to a merely animal one? No, not even that.

What is this: "Decision for the total disvalue of life"?

The examination of the problems of existence from the aspect of the individual. The individual only being in the community. Existence of the community of people, the coexistence of people, and so on.

Husserliana 42, Appendix 46: Type and ideal model of a personality. Exemplar. Love. <Normal types as ideals of self-perfection and of the evaluation of others.>[3]

The Greek—the other human beings and peoples, "barbarians." The Europeans—the other human beings, races, peoples, "lower" races.

In European supra-nationality, each nation, which is national in its constant national self-consciousness as a special form of humanness, in that they know themselves as English, as French, and so on, and know themselves as an <excellence> for each nation as an exceptional and excellent value for it. What does this mean upon closer examination? What kind of excellence? How does someone who is French see their being French as an excellence, the English, the German, in their being English or their being German, a value common only to their fellow people, and yet personally and responsibly belonging to each individual, a value that no one may or can—even if he could, he would not—exchange for a value (be it also one recognized as objectively the same) peculiar to another, to other human beings of other people? That belongs with the whole title "love for one's own nation," under the broader title "love"—in an excellent, though in very general sense: love for people, for socialities, for nations, for the supra-national, humanity—modes of neighborly love. Furthermore, as an object of love, social institutions in their function, in

[3] <*Husserliana 42*, 523–526.>

general all the manner of personal existence, personal striving, personal ends that can even be themes of love. [524] In addition and correlatively: love for the personal in the worldly. One loves the person, one loves their personal dwelling, the one furnished by them for personal use, the dwelling arranged in a homey way, their house, their garden, and so on. Likewise for a nation, the national homeland, "fatherland" as fatherland, the cultivated landscape, and so on; "love" for a dish, as a mere desire for it is not yet love. If I love a person, however, I can thus have an indirect loving interest in their having this kind of dish, this kind of taste in terms of sensible nutrition, and so on.

Love in the genuine sense is one of the main problems of phenomenology, and not in abstract particularity and isolation, but as a universal problem—concerning the intentional elementary sources and concerning their disclosed forms of effective intentionality springing forth from the depths to the heights and to the universal expanses.

Regarding the form of love, "nationality," the following should be noted in advance: everyone as a member of his nation (and of every relatively closed and generative sociality, family, tribe or village community, and so on) knows themselves as that, namely, as a person of a certain normal type "French," "English," and the like with corresponding associated types of anomalies, both constituting in the unity of relatedness to one another a type. But with such a nationally typical human existence there is a loving aspiration towards a national personal ideal (of course, differentiated according to the difference typology "man" and "woman," also to the typology of age that belongs to the very general human typology). As a human person living in the spiritual space of national sociality and in the spiritual orientation form (I in my family, from there in the circle of my acquaintances, and in the movement of opposites, through which I constantly gain new acquaintances as living in unity "with" me through socialization), at the same time as I strive for the increase in fulfillment, towards the ideal, I also strive for the correlative social esteem, namely, to be seen and to be esteemed by them as a relative realization of the ideal.

Obviously, this is not so easy to accept. We have the differences of the status, the class structure, the village, country division; each has its general personal typology, and each one has its particular ideality, its direction in the striving for self-perfection. An ideal French or German—in relation to the historical times, thus, at first, our time—indicates a relatively realized personal ideal, but each in its special national space, in its living, generative and relatively closed sociality having its own relative historicity. [525] According to what was said above, the ideal is not only an ideal seen from the outside, a *limes* for the increase, and, moreover, a value, but the perfect Frenchman in question has consciously become as toward the ideal for themselves, that is, as through the will to self-perfection. Only that, in lovingly living towards the ideal, can he be aware of how far behind it he is, while the other, as co-living and co-loving the same in another way can also critically notice such things. On both sides, there is a loving idealization in another sense, namely, the overestimation of the approximation to the ideal with regard to the friend, in the love between friends, of the beloved woman in sexual love, and so on; there is an imaginative overestimation of the realization of the ideal promoted by this love from person to person and an underestimation, that is, neglect of the *negativa* and not just incremental changes.

Similar differences between the mere normality of the type and the normality of the ideal value correlative to love in everything that has spiritual sense from persons and for persons: national environing world, national poetry, the work of poets themselves and the poets who have been perfected and objectified themselves on the side of poetry in poems that have become objects of general love and realized ideals.

Loving joy in children—not in one's own children—in a striving to educate them to the ideality that lies within them. Loving consideration of humans—love of neighbor—not of humans with whom we are habitually and personally "bonded" in love.

The ideal in me as my exemplar, I as the exemplar of myself. Others, reverenced, beloved people as my exemplars; living in accordance with them, I want to become more perfect. What is here the first in itself? Are there not, if a deeper interpretation is attempted, some slight problems here? Is exemplariness "imitation" ("*imitatio Christi*")? Great, excellent people have those who ape them. Is it not in addition to the imitation? Yes, because it is only directed to externalities, one may say, while such aping ones are not able to understand the spiritual greatness, that which is properly worthy of reverence.

Nevertheless, has everything been settled? Is that life in the motivation of exemplariness an imitation? An aping-after, even if reaching into the deeper levels? A mimicking? Being human is being in activities, in actions, and in habitualities from actions. One can, to a large extent, mimic everyday and habitual activities, lived-bodily actions, and more mediate actions, but a person in their habitualities cannot be mimicked. In a mutual understanding with another person, in their now-being-motivated-in-such-a-way-and-living-accordingly, in their personal historicity, I am, and possibly lovingly, in coincidence with them, I live to a certain extent in them, in their development, in their now-life.

Husserliana 42, Appendix 47: <Universal reflections on the entire value of human existence in the face of irrational fate and total practical doubt.>[4]

The naive practical existence of the human being in "practical positivity." Motivations of naive existence with regard to this positivity, with regard to the belief in the possibility of an existence worth living, that is, for these to become doubtful in a world that enables a life worth living in it. Do not such motivations belong as possibilities to the general human existence? Does not a universal horizon of fate belong to the specifically human existence? In other words, does not *practical irrationality* itself belong to the essence of practical life, as a life that presupposes "rationality" but, in

4 <*Husserliana* 42, 526–527.>

its rationality, is constantly threatened through irrationality, in the *possibility of existence* that is to be maintained? And does not the constant possibility of suicide belong to the practical horizon of human life? Do not, therefore, *motives for universal reflection on the entire value of human existence* belong to the essence of human existence, of whether it can be of positive value at all and how it can become so practically from human power of will? Universal reflection of the moment: in despair. What can I still do? For what is left for me to hope? Can I still live? Is voluntary death the right thing? Or full renunciation of the world in the manner of Buddhist release? Likewise in empathy for the other.

But in the general experience of the fateful existence of all people, there belongs to life the *turn to the general question*: *What can we humans do at all in the face of fate* and the *possibility of situations of despair* that we all face, or total practical doubt, that call into question the totality of practical possibilities? In them, the "possibility of living" becomes questionable, all of human existence as such becomes questionable in its "possibility," [527] that is, questionable in the possibility of a satisfying affirming existence in worldly life in the form of possible satisfaction. Satisfaction is satisfaction of the will, but in the universality of progressing will. But in despair: How can we hope to overcome fate, despair and regain positivity to be enduringly grounded through one's own deed? Universal self-reflection, universal reflection of human practical existence in general, reflection on one's practical possibilities in the world in which one lives, in which one lives as one's practical field in everyday naivety. The original instinctive and traditional positivity as a constant drive towards positivity. *Religious belief* as an originally naively emerging form of overcoming, for which the drive itself originally inclines.

If universal "theoretical interest" has already awakened in humanity as practical philosophy, it can take the form of an *attitude of universal, theoretical reflection*. The one in the theoretical attitude becomes the detached spectator; one brackets one's interest, one's being "interested" as fate-conditioned person in being and not being; the theorist is interested in being and not being in the sense of truth. One becomes a scientist, whose theme is humanity, fate, the world of practical people, and so on. But that leads to universal science as such and initially to universal world science.

Chapter 5
Husserliana 14, No. 9: Common spirit I. – Person, Personal Whole, Personal Effective Community, Community—Society.

Edmund Husserl

1) *Social acts and how communities are instituted by social acts; especially: personalities of a higher order. Different forms of communities and higher-level personalities. Naturally formed communities and their development (family)—"artificial," arbitrarily instituted communities.*

2) *Ethics and sociality (in detail). Duty, responsibility, functionary of a sociality.*

<§ 1. The subject of drives before and after empathy.>

If we imagine a subject before empathy, constituting a spatial world by way of drives, this is not a *person*; even if we add to this the freedom of deliberation, the suspension of driving realization, deliberating about the best, the best goals of pleasure or the best ways (material pleasure and sensible displeasure, pleasure-things, and so on). That is "reason" at this level, and there are "objects of reason," namely, technical objects, tools, "methods" "thought out" to bring one to delight, to flee danger (originating from things), which are transferred from one case to new ones and thus become apperceptively determined methodical apprehensions.

Up to this point, we have not taken into account other subjects as targets of effect, as subjects who "do us good" or "do us evil." I can let myself be determined by others:

<Edmund Husserl, *Husserliana 14. Zur Phänomenologie der Intersubjektivität. Texte aus dem Nachlass. Zweiter Teil. 1921-28.* ed. Iso Kern. (The Hague, Netherlands: Martinus Nijhoff, 1973), 165–184.> <Freiburg, 1921.>

E. Husserl
Freiburg im Breisgau, Germany

S. Heinämaa, A. J. Steinbock (eds.), *Values of Love and Ethical Reflection*,
Husserlian Legacies: Themes for the 21st Century,
https://doi.org/10.1007/978-3-031-68698-6_5

in the manner of *imitation*. The I can by way of *drives* want to help them. [166] There is a "motherly-love" that is of drives, a "parental-love," a care that is at the same time in its fulfillment a shared joy in their welfare; of course, it is also directly a compassion with their suffering that is of drives. Strictly speaking, it is not a shared joy in their joy in the sense of rejoicing in the same thing they rejoice in, but a joy that they rejoice, a sorrow that they are sorrowful. But also a love for *them*, a joy in their existence, a joy in living together with them, a rejoicing in their perceptual existence, in them as they are given in empathetic experience, in their egoic living and doing—a general joy that does not exclude taking exception to particular deeds, evaluating them negatively, intervening against them, and so on. Joy in the subject as identical subject of the comportment toward the environing world, showing a specific character of its own in the manner of this comportment. Love determines an interest in support of the other, hatred an interest in harming and annihilating the other.

These are still not *social acts* and not acts of authentic social love.

<§2.> Social acts and relations. <The I-You relationship.>

1) I do something with the intention that the other notices it, and in the expectation (thus also with the intention) that the other behaves in such and such a way. But that does not yet make it a social act. The other need not notice my intention (I do not *turn* to him).

2) I do it with the expectation that the other, noticing that I have this intention, does something. Am I thereby already turning to him? I turn to him, first and foremost, when I have the intention of *communicating*. I announce something to him; if my wife puts an apple on my hat so that I remember to eat something before leaving, I understand her intention. Is that communication? It is certainly an announcement of her intention. Another example: travelers who put branches on the crossroads so that their companions may know the way they have gone—one must address this as communication. [167]

Immediate communication, or better, *contact, establishing an original nexus between I and You*, in the original empathetic experience: We have the original experience of a facing-one-another, and I "say" something to him, I "express myself"; I perform an expressive movement or a vocal utterance or perform an outwardly visible, noticeable deed, which is capable of awakening in the other person the consciousness that I have the intention to announce something to him; in doing so, the expression of this intention (as with calling to someone) can be different from what is now to be further announced and communicated. More exactly, what now follows, this is indicated as "that is so and so," "that, you should do" and the like. The communicated matter of fact is the content of my intention *that* the other takes note of it, and indeed, of my intention announced to him. The form of the communication is understandable through the preceding call, for example, and possibly also by the linguistic form to which this form of communication adheres as such.

How is the *I-You relationship*, which is the personal effective relationship in the concise sense, established?

1) Speaking metaphorically: Both of us, I and thou, "see eye to eye," you understand me, maintain me, I maintain you, simultaneously.

2) I turn to you and communicate a matter of fact to you: I experience a matter of fact and draw the attention of the other, in whose proximate sphere of experience it is also, to it by a "pointing." A certain hand and finger movement, throwing a piece of wood in the direction, and so on, awakens attention, "steers" it in that direction in which something interesting for the other (as for me before) takes place, and the interest naturally, by itself, glides away from the pointing movement to the new thing to which I wanted to call attention. Such means of arousing an interest in a present objectlike formation in the external surroundings of the I-You are, if they often present themselves as suitable in the same way, used often by both sides and are finally understood, without further ado, as produced by the other for the intent of "pointing out." Thus, in the given case, the You understands: "I see this and that there and point it out to the Thou so that he also sees it." My intent that he sees, my will, is realized by the means of pointing out [168] (pointing out <as> the way), which is for me an action that he is supposed to grasp, to understand as a means by which I want to steer him towards something. I thus have an effect on the other, and not only in general, but within the framework of the reciprocally and simultaneously performing empathy that brings I and You into "contact"; and, moreover, doing so in the particular way that my effecting (that is, my will or my instinctual striving to "determine" the other, to "move" them spiritually to determine them to a striving or wanting) should be noticeable to the other and itself belongs to the "way," to the means of the determination.

Normally, this is an effective means: of course, it is not always so, as in the case of the fear of the enemy, of the flight that may follow, and so on.

Another case is descriptive communication (in the situation of "contact"): expressions, natural-bodily reflexes of external processes, of external physical or animal occurrences that become natural indications of these processes themselves, drawing the attention of the partner, if he has not seen the processes themselves, then serve, produced with an intent and emphasized before the other to whom one makes them especially visible, serve as something pointed out, and at the same time as particular, for particular sides or parts of the process.

Then communication of recollections and of that which is not present. Furthermore: communication to ones absent, or rather, of ones absent, that is, to others temporally distant. My present willing here concerns a future understanding of my sensorily persisting communicating expression and a future taking note of the other (thus, a comportment of doing something), namely, in a "later" future lying outside of the sphere of the present.

The I and You do not "contact" each other. They reach out their spiritual hands to each other over some span of time, the past I is the subject of a communicating act, is the giving, the later, future one, the receiving. But it also belongs to the factual situation that the I, insofar as it now wills in such a way, thereby institutes a willing-sentiment, a habitual, enduring will and "remains," at least in general, with their will. And so it will be understood. The receiver takes the giver as a human being who is not present, but as one living in the present who wanted to give him this communication then, and *still wants* them to receive it; [169] this is so if no countermotives speak in favor of a modification of willing, of sentiment. We also note here that, at the same time, one who is deceased, and is recognized as such, and a living person reach out to each other spiritually; that a communication can come from the one to the other, understood as addressed by the once living to and received by the now living.

<§3. The practical community of willing.>

We have focused so far upon the mere communication of facts; a certain community of willing and a certain understanding is already present here. But there is another, in a concise sense, *practical understanding, a practical community of willing*, namely, that in the state of contact the Thou can be determined not merely to take note, but to perform another, further action, for example, an external action, which has an effect on the physical or spiritual environing world. The one who is addressing announces and possibly descriptively communicates the wish or will that such and such occur, and in the expectation that it will motivate the one who is addressed. The situation is similar to that of mere communication, only that this becomes a means for the execution of a further action (in contrast to the taking note that is in the communication, which is also already an action).

If he does not want to I can want to force him and count on the fact that he knows it, or I can also communicate this to him, and he can, just like in other cases in which he avoids external circumstances that bring pain, act according, he can submit to my will—my will communicated to him is possibly the motive of his enduring, general will to do once and for all what I want "in general," just as I can want him to comply in general, insofar as I presume that I at all possess the means of coercion. Then the relation between master and servant arises as an enduring personal relation, as enduring through a universal habitus of persons who had entered into contact or mediate spiritual reaching out to one another. [170]

If the relationship is instituted, then every action in which it appears is characterized as emerging from the instituted interweaving of the wills of *both* the persons. I command—I as master, he follows "dutifully" as servant, in the consciousness of having subjected oneself, of being subjected, and with regard to acting in the consciousness of his dutifulness.

In the *request*, we have a temporary relationship. The communication of my wish will, I hope, act as a motive for their will, my wish and my hope that its being noted by the addressee <will move him to comply> are directed to them. The wish concerns their striving, willing, doing; thus, it is not yet directed to them, but it is so through the communication, through the taking note performed with it, and in the contact of I and you.

And instead of immediate contact, a remote determination can always occur, whereby the communication itself can take place in the contact of the subjects as in the case of a present wishing and willing directed to the future, or the communication itself can be a remote communication.

Community of wills, mutual understanding can then also be *reciprocal*, ending in a reciprocal agreement. I fulfill your wish if you fulfill mine; I grant you this, if you grant me that in return. Furthermore: We both wish that something happens, we decide "together," I do this part of it, you do the other part. And so on. S_1 and S_2 want the same G, but not each for themselves, rather S_1 wants G as that which is also wanted by S_1, that which S_1 wants also belongs to the will of S_2 and vice versa. That S_1 realizes the part D_1 and S_2 realizes D_2, that again is included in the will of both, and is included for both as a "means" (in the broader sense) or as belonging to the realization, but before to the intention.

<§ 4. The gaining of personal self-consciousness in the I-You relationship.>

The pole of all affections and actions that is necessarily present in the solipsistic, fictitious subject, that is, the subject of motivation passing through the stream of experience, which is, as such, the constant subject of a striving in manifold modalities, becomes the I. [171] Therefore, the *personal subject*, gains *personal "self-consciousness*" in the I-You relationship, in the community of striving and community of willing made possible by communication. In empathy, the I already becomes *aware* of itself as subject of its life and subject of its environing world, and is conscious of the alien I as an "other I" (but, that is, initially only as another subject pole of another life that is related to the same environing world in other manners of givenness, and so on) through the reflexive relation to its own, thusly reflecting subject. I as subject of motivation not enter into the primordial-social I-You relationship not only alongside the other as other, but I motivate you, you motivate me; and in the preeminent relationship that the I-You relationship establishes through social acts, there exists a unity of striving or specific willing that embraces both subjects; in this unity, both are mutually related to each other in the present consciousness of mutually "exercising an effect" on each other as striving subjects, that is, in being directed towards each other in striving, they determine each other to a doing or to an undergoing. In this community, not only does everyone strive, but each is also to themselves as an object, as the one striving; one is not only, as that, pregiven to oneself, but given to oneself as an object. Of course, not as an object as a theoretical theme, as a theme taken up into attention and an intention directed towards closer attention or even theoretical knowledge (attentive examination, observation, knowledge-seeking and knowledge-acquiring perception); but as an object in a practical way, namely, belonging to the practical theme. At one time, the striving goes to the appropriation of the object as a "thing," to its exhibition, to the production of the ideal identical unity of the intended, to the transformation of the intended in its self that would be grasped as identical in always newly intended productions, possibly then further to conceptual grasping, predicative determination, to *theoremata* pertaining to the thing. At another, the striving is directed toward "moving" the other, that he fulfills my request, my command, first of all, that he takes notice of such and such to which the request, the command refers; then, that he takes notice of this, my command, my request; and that my <command>, [172] which has been given to him, the wish of which he has taken notice, should "move" him accordingly, be motivating. Here, then, I myself belong to the practical theme, in the context of that toward which the practical intention is directed, in the context of that which is to be practically surveyed and foreseen. Thus, for every action, I must have the "way," the middle goals, and the final goal practically thematically in mind.

<§5. Love.>

Personal love—an enduring sentiment, an enduring practical habitus, its realizing operation:

1) An active pleasure in the personal individuality of the beloved, in their whole bearing, in their passive and active comportment to their environing world, in the expression of their individuality in the lived-body, in the thoroughly spiritual lived-bodiliness in general.

2) A striving not only for the richest possible operation of this joy, but for personal "contact" with the beloved and for community in living and striving, in which their life is taken up into my life, namely, their striving into my striving, insofar as my striving, my wanting is itself realized in theirs and in its realizing doing, as theirs is in mine. The I is the pole—for affections and actions, as that, it is a pole of a unity of *striving*, going throughout the whole streaming consciousness. Striving in its manifold modalities makes up the life of the ego; all intentionality, which <makes up> the form of spiritual possession, of freely available, acquired unities of existence, and of valuable or useful existence, is a structure of striving and possibly of "conscious" volition. But if I come into a community of striving with another, then I live as I in them and them in me. But the kind and intimacy of this community depends on the extent of the security in one another of I and You, on the scope of the community of striving and, further, on various other points. Not every community is a reciprocal community, and not every reciprocal community is a community of love.

Idealiter, it belongs to the community of love that those who love each other as personal subjects, [173] that is, as socially communicating striving subjects, have not only instituted a community of striving in individual directions of striving, but that such a community is instituted in an authentic and also inauthentic manner for all of the possible striving of both: the lovers do not merely arrange to meet for some special common undertaking, which of course may also occur, among other things;

before that: they do not merely reciprocally communicate with each other, do not <merely> engage in a community of taking notice with regard to their common surroundings, and so on, but rather, included in the fact that they have been bound together in a community of love is that, in a universal way, all striving of the one enters, that is, has entered once and for all, into the striving of the other and vice versa. Included in it is that, as soon as the contact of the lovers is actualized and one becomes aware of the striving of the other, he *explicite* ("authentically" so) seeks to take up the striving into their own, possibly with the participation of the will of the other who might intervene helpfully. And if the other prefers to finish matters themselves, still, what the other does of their own accord is not only their business, but also the business of the lover in whom their wish and will is also realized: not willing just for his sake, theirs as an isolated I, but for the sake of the other, because this other, as the subject of their living and striving, is completely taken up into the scope of the striving intentionality of the lover. Even in the operations of living of each of the lovers, which take place outside of contact, of which, to a great extent, the respective other knows nothing at all, the will of the other lives *implicite*. As a lover I know that whatever I think, feel, strive, do, all this is necessarily "in the sense" of my beloved, is right for them, is not only not only not blamed by them and, in this sense, recognized as right, but, as striven for by me, also is in the sense of their striving. This sense would be actualized, if they were there, would participate with joy, would sincerely like to stand by me, and where this would be excluded or not wished for: the beloved would gladly and lovingly put themselves into my striving situation and, in willing approval, would know their I and its striving to operate in mine (inauthentic, implicit life and striving of the lover in the beloved). [174] It can be said: lovers do not live next to one another and with one another, but in one another, actually and potentially. Thus, they also share all responsibilities, they are bound together in solidarity, even in sin and guilt.

<§6. Ethical love.>

But the question is whether this love is not a limit case; it makes up the concept of love in each good sense. This described love is perhaps a sinful love, or includes all sinful love within it. We are thinking here, of course, of Christ's infinite love for all humans and of the general love of humanity that the Christian must awaken in themselves and without which they cannot be a true Christian. There was no talk of ethics in our description of that love. Loving parents chastise their children, and they do not take up all the living and striving of the child into their own living and striving in the manner described above. The friend is grieved when the friend falls away from his "true self." The Christian who practices love of the enemy does not love the evil in the enemy, and the evil deed is not that of which he approves in his will. In every human soul lies—this is the faith—a vocation, a germ that is to be developed automatically, for the good. In everyone, there is included an ideal I, the "true" I of the person that is realized only in the "good" action. Every awakened human being (the ethically awakened one) willingly sets in themselves the ideal I as an "infinite task." And this, then, results in special ethical forms of community, depending on whether, of those who are bound together, both are not ethically awakened, or one is ethically awakened and the other is not, or both are ethically awakened. And then similarly in the case of a plurality of personal subjects. Thus, as a truly (ethical) lover, I love and I live willingly along with the other in their germinating soul, in their germinating, growing, becoming ethical subject; or I live in the maturely awakened, wrestling and struggling subject, freely ethically being worked out in all that arises from the other's positive ethical habituality as their ethical personal life (the negatively ethical is against my wish and will). I live in it, at first, I approve, I rejoice (or deceive myself). [175] But ethical community is ethical friendship, ethical relationship between Christians, and so forth.

<§7. Love and community of love.>

Love of neighbor, which consists only in loving care for the other, for their ethical being and becoming, is a sentiment for others, for communities, for the whole of humanity, but is not a communal relation, not a personal bond, nor even instituting a personal association.

We will therefore have to distinguish *love* from the *loving community*, the loving interpenetration of otherwise distinct persons into a communal personality.

Christian love is, at first, of necessity, mere love. But it is combined with the striving (which is necessarily motivated by love) to become a loving community to the greatest possible extent. In other words, striving to "enter into a relationship" with human beings, to open up to them and to have them open up to oneself, and so on, all according to practical possibility, the limits of which are set ethically and thus by ethical love itself.

<§8. The origin of personality in sociality. The community of delight, care, the common meal and the family.>

Prior to the ethical are the personal manners of comportment that are subject to ethical assessment, to approval and to disapproval. All such manners of comportment are not manners of comportment of a subject in general (as some jellyfish subject), but precisely of a personal subject. *The origin of personality* lies in empathy and in continually growing *social acts*. It is not enough for the constitution of the personality that the subject becomes aware of itself as the pole of its acts; it is constituted only when the subject enters into social relations with other subjects, whereby it already becomes practically objective, as we have shown. In this way, it becomes a subject that can and does enter into community, [176] but that also enters into occasional personal relationships with others and now exercises, in its living and striving, not only self-preservation vis-à-vis the world of matters, but also as a person in the world of persons.

Each person has a thing-world, a mere nature as surroundings. If we screen off all personal relationships, interests, strivings, the question arises as to what remains in terms of values and practical goals. Sensible desire and corresponding strivings and willings. Is striving for observational taking notice of and for theoretical knowledge not also conceivable; joy in beautiful appearances, in beautiful nature, in the examination of pictures of nature, especially those that exhibit beautiful manners of appearance, joy in the formation of that which is beautiful?

What characterizes the sphere of naturally healthy sensual pleasure and normal, healthy sensual striving, the transgression of which is announced in the consciousness of the abnormal, the unhealthy, the degrading, the mean? Is this an original distinction or a derivative one? Can scientific and aesthetic intention be motivated in this screening off and establish an ordered rank of values?

Let us cancel the screening off: I not only have a physical environing world, I find in it bodies and other subjects, at one in lived-bodiliness, I find animals and humans.

Which possible relations arise in relation to alien animal and human subjectivity? Which possible valuations, value meanings, strivings, willing goals? Which possible connections as personal, to personal wholes? I can remain with the body of the alien animal and human and evaluate it purely as an object of physical delight. That is one thing. The subjective is then not attended to, though it is there for me. Just as when the body is killed, eaten, and so on, for sensible delight.

But the soul and the subject are there. And the question is how its existence motivates me. I must, acting on the lived-bodiliness, determine the soul and can cause sensual pleasure or sensual suffering in it. I can, treating the alien body as an object of delight, thereby also cause pleasure in the alien subjectivity, or displeasure, pain. [177] As in sexual delight, not only this can take place, but both can be in the delight, not only each for themselves, but there for each other as delighting with and through each other and as striving for delightful activity at the same time as they are striving for the delight in this entanglement producing a *unity of delight of the community*. The other and their consent, or at least compliance, is not only the means of delight (that is, not only the alien body, but the alien subject as their lived-bodiliness, which lies in their sphere of power, offering and allowing the operation of pleasure), but *that* the other participates in an operation, that a unity of will encompasses both and brings about a unity of mutual action and that both at once delight in pleasure that is itself the object of delight, for each, that is, that the other delights.

But it can also be that no consideration is given to the alien subjectivity, that the delight is imposed against its will, even that it is annihilated, and that suffering is thereby imposed upon it. Then the higher layer of joy and of value not only falls away, but a negative one takes its place, of which it is to be asked whether it does not bring about a conflict of values, which not only diminishes the value, but cancels it. Then the following comes into question: the coercion to will (thus submission of the will of the other), and the forcing of the delight under counter-will of the other, without even forced consent, submission of will taking place. It can be that, on the coerced side, no wish existed, but in the submission, pleasure arises and the wish awakes and, thereafter, satisfaction. It can be that no pleasure takes place, but rather suffering, which is accepted to avoid greater suffering. It can be that suffering is endured (put up with), but not "accepted wholesale"; finally, that suffering is not even endured, but rather striven against in persistent rebellion. To be sure, endurance is also counter-striving, but at the same time it also includes an acceptance in the heart that is missing in the unbroken and unyielding rebellion. Likewise, for bodily pleasure, which is biologically connected with reproduction and preservation of the species, but does not concern the act in and of itself [178] (in animals, the smelling, and so on, the licking of the young, and so on); the compulsory washing, cleaning, restraining, pulling away of the children by those rearing them, whereby trouble, suffering is inflicted on them.

Instinctive care for the lived-body that is in service of the healthy preservation and of course the psychic well-being in advance: in humans instinctively and yet also with premeditated intention. As the child develops, they learn to understand the demand, they learn the ought of cleanliness, and so on, they learn to submit, to be patient, to obey, and, finally, they learn to want freely and to strive passively of their own accord for what was previously forced upon them, they learn to understand for themselves the value of health and to resist for its sake a momentary stimulus, a momentary comfort, convenience.

Comportment to children who are "not fully" developed, who are considered, valued, loved from the point of view of future development.

Comportment to animals as unreasonable, beings of drives and to higher animals approaching the level of humans. Developed humans in relation to developed humans.

Temporary relation of communication.

Instinctive striving for being together with immediate members of sexual reproduction (family), commonality in sociability. Common meal. We assume that everyone can satiate themselves fully and undisturbed by the other, motives of envy, and so on, remain switched off. The joy of one's own food, but also the joyful resonance of the other's joy of eating. On the ground of the natural love for the family members. This is a temporary community. But it becomes a regular institution, regular coming-together for a common meal as a unity of social action. The common meal is an element of the institution of the family as a social institution, as a lasting community; as that constituted for the consciousness of the family members. The family is not only a habitual kind of living next to each other, remaining near each other in life, but a community of life with living arrangements that have a social character. [179] That is, each member not only *knows* the other next to them and not only seeks the other because it gives them joy to be near, but rather there are commonalities, mutual interdependencies of striving life, of doing, of interacting, of functioning-in-relation-to-each-other, whereby the doing of the one enters into the doing of the other; in this way, each subject intervenes in a personal manner in the living and doing of the other, thus living, uniting as a person in manifold I-You relationships, in short, a unity, a whole is constituted, built of I-subjects that interpenetrate each other in a connecting way, insofar as the life of one I lives in the life of the other, participates in it: the I-hood of one is not next to that of the other, but lives and acts in the other. In the family, this happens in all the I-relationships that institute unity in the family, which do not encompass all the relationships that can put human beings in unity with human beings, that can bind them with each other. The father of the house can have the social function of the master, the other family members the function of obeying, serving. He can be a despot or he can be the "head" who cares responsibly for the whole. And the mother of the house can be, as woman of the house that toggles tasks, the head of housekeeping and the care of the children, responsible for it, and can be, in this function, of lady, on the other hand being, nevertheless, accountable for the correctness of her actions to the master, who is responsible for the universal whole. In this way, the other adult members can have their special functions and act responsibly in them, again under a designated restriction: they may be under the woman of the house, as economic functionaries, or directly under the lord as the supreme "head." Everyone knows themselves and everyone else as subject of their own functions.

Naturally-instinctively growing personal associations (family and those originating from the family), *historically* transformed. [180]

Artificially instituted for some purpose, clubs, associations.

<§9. Sociality and ethics. The function in the community and duty. Community and society.>

Reason in sociality. The Ethical. The ethical on an instinctive basis. The ethical and personality. Can there be an ethical where real or possible personal community is not involved, some comportment of persons to persons?

But there, *personal relations to oneself* come essentially into examination. In the constituted association, I am not only active subject, but subject of a *personal action*; as such I am not only subject of the praxis, but also object-subject, *belonging to the content of the action* (the willed, striven for, planned). If I make an agreement with another, if I take on a function, then the other determines me to determine myself to take on this or that. In the naturally growing family community, we easily see that the first is the natural, naive growing care of the mother for the children, of the husband for the mother as wife and as mother of the children, and so on.

The omission of natural care by overlooking, by momentary egoism, by way of unreason, and so on, leads to criticism, then to *personal injunction*, to the command, and so on. *The ought arises*: the "he ought" and on the part of the one who experiences and takes on the demand, the "I ought." There is no lack of motives of "I ought," or rather, beforehand, of self-assessment, for the father of the house either he becomes sensitive to his own failure, for example, and the wrongly given order harms his relatives and himself. The whole action, in which he himself as the leader was and, in the memory, is objective, is blameworthy, and he is its originator. He decides to give other, better orders that have the character of not only better but of *ought*.

Each member of the family is a responsible subject, a subject who has a generally circumscribed and, in the particular case, determined I-ought that belongs to this generality. This enduring (*habitual*) I-ought designates here the naturally arising duty, [181] and the same duty has a general content that generally circumscribes the dutiful as the sense content of the duty, as in the particular case the determined particular duty. If we say that father, mother, and so on, have these and those functions, we conceive of the family as a personal union which bears in itself a common value as an end that is the end for each family member, while, at the same time, each has their particular ends included in the entirety of the whole end. Each member has the function of helping realize the familial end by realizing their particular ends, and to that extent it also has this function, it is the relevant subject of ends. The ends are, of course, *generally* circumscribed, and that is what the functional expressions—father, son, and so on—denote. Everyone has their own function and, as a consequence, their duties to fulfill their functions, and this in as perfect a manner as possible.

Is there an essential distinction between function and duty (of ought)? *Function* refers to the practical determination of the subject, the ordering to an end, namely from the point of view of a special end that serves a comprehensive end of the entire social association, here the family. The family itself has no function in and of itself, but only, for instance, in the life of the people and the like. But in the family the father has the function of the head, the woman the function of the wife and mother, and so on. In the expressions "*duty*" and "*ought,*" the passage through the negative resonates: the deviation disturbs the concordance of the will and causes the reaction of censure.

In "*artificially*" *instituted communities* of coordination and subordination (master and servant, club, building contractors), the functions, or rather, duties are taken over voluntarily on the basis of the agreement or, as in the slave relationship, imposed by coercion on the subordinates. But a willing connection, a willing unanimity also makes here the personal connection. It is actually broken if the members refuse to fulfill their duty (regardless of whether good or bad, whether happily or unhappily and with nonchalance). The runaway slave is no longer really a slave (apart from the question of law, which does not yet play a role here [182]). A slave revolt breaks the master-slave relationship. The coercion, the suppression of the uprising, restores it; and the expression fits, provided that the new relationship becomes conscious as a continuation of the old. The slave recognizes the ought, and also the ought retrospectively for the time it was broken.

Changes of naturally arising personal communities (or personalities of higher order) by partial transformation into a community of willing. In the state: a community of willing with respect to all full citizens. Those who are not yet enfranchised, those who do not yet participate in the functions of the state, form a host of adolescents, when appropriately mature, subordinating themselves and, in this way, belonging to the community of willing. But they cannot, through their will, contribute to the state constitution or its change. But this is insufficient and requires own considerations.

Cf. Tönnies[1] on the terms *community* and *societies*, who does not conceive of community as a community of willing (will understood in my concise sense). Linguistic community, community of family, marital community, community of the people. A linguistic community is not a personal bond that creates a personal whole, but a marriage is, even if it is a "modern" marriage. A *state* is a personal whole, though not everyone knows each other, just as in a larger club. The way in which a personal bond *is established* must, of course, *start* from direct empathy and direct agreement or from naturally growing subordination, and so on, which is instituted in the state of personal contact or communication. Then, however, the institution of personal associations must be considered in a *mediate* way, wherein the "persons" remain "unknown." They are, however, in any case communities of willing of definite persons who are therefore in understanding as subjects of will, even if mediated.

A linguistic community is not a personal bond to a whole which is based on unity of a purpose arising arbitrarily or from "natural conditions," and which has been bound by the unity of an end and by the coordination of function and duty. [183] Likewise, for a community of a certain social position, a community established by unity of a custom, a form of education, and so on. Finally, the unity of a people, a race, and so on. If the people is a people of a state, the state unity and the unity of the community constituting forms of custom, and so on, are nevertheless distinct. We include race here provided that the commonality of the external physical habitus goes hand in hand with such community character. Otherwise, it does not belong here.

Community does not mean uniformity of kinds, forms of personal actions, of ways of thinking, opinions, scientific activities, and so on, but in community stand persons who in this respect stand in the unity of a spiritual connection of effects, whether, in the individual, the effect becomes visible everywhere or not. In the broadest sense, in the unity of a "tradition," in a personal wholeness, in a personality of a higher order, I act on You. The will of the affecting passes through the Thou, as a personal affecting, and in the functionary the personal will of all other functionaries effects *implicite*. *The linguistic companion*, however, has an effect far into the linguistic community by means of a coined expression that "finds appeal," *but not personally*. In contrast, the *poet* addresses his audience with his linguistically coined work of art, similar to the actor, like the teacher addresses his students, like the scientist addresses his fellow scientists. The community of scientific researchers is more than something like a linguistic community. Insofar as the researchers think of their *co-researchers* (indeterminately co-imagined of as a universe, but possibly also of the familiar ones "belonging" to the present generation), turn to them spiritually and know them as the ones to whom they turn, they form a personal bond, but certainly not an association that has its constitution.

The community between the artist and his audience, which follows him or rejects him and which he has before his spiritual eyes as an indeterminate generality, is,

[1] <Ferdinand Tönnies (1855–1936), a German sociologist.>

once again, a different matter. Furthermore, the merchant and craftsman and their real and possible clientele. There are obviously personal relations here that go from I to I, but no club, no personality of a higher order, the concept of which would now have to be determined in more detail. [184] A speaker's audience: most of them may have fallen into it by chance. But the speaker can also have a more or less regular audience as a professional. Just as the craftsman, merchant, and so on, has their casual buyers and yet a regular clientele, which also includes the average number of coincidental buyers, who change, but according to number have under given circumstances their regularity. All such things are to be thought through.

Husserliana 14, Appendix 23: Influence of ego-subjects on one another.[2]

1) In understanding the other, I have an effect on him: he has need, I remove it, he is hungry and I prepare food for him; and he himself need not know anything about it, and he does not ask, and so on. All effect of I on I presupposes empathy and is constantly carried out in empathy or in the mediate re-understanding that follows and is based on it.

The following case does not belong here: A is sick, B, recognizing this, avoids him. This is a special case. I also avoid stinging nettles, but I understand the illness of the other person through empathy with their lived-bodiliness and through the knowledge that such illness is transmitted. Here, however, there is no influence of the I upon the I, and we only have such cases in mind. The influence of one animality upon another that is not reciprocal egoic act-motivation, not the one ego determining the other.

But the following does belong here: A wants to do something evil to B, he may conceal it, but B sees through it and he flees or takes precautions. A can understand this again and, for his own part, acts accordingly. A and B need not directly encounter each other here. A can also have some indirect indication that B has this and that intentions with respect to A.

2) *Suggestion*: For example, I do something, the other copies it. This can lead to willful influencing of the other person. I know that the other person likes to copy what I do, and I want it to be so in a particular case. So I do this and that for him to see. We have here, then, two cases of suggestive effect: a) A does something and B is thereby suggestively influenced (perhaps wholly unknowingly); b) A knows that B (and that others in general) can be suggestively influenced (before all intention), and this knowledge motivates a volitional act of A to influence, by means of suggestion, the will and action of B in a certain way. [185]

II. From all these cases are to be distinguished those of reciprocal action through *social acts*. I direct my will, my "I want" to the "I want" of the other in such a way

[2] <*Husserliana 14*, 184–185.> <From about 1920.>

that I ask them, command them, direct myself to him and his actions, in the way of mutual I-contact. I not only know that they will act in such and such a manner, that they will let themselves be motivated by me, and not only that I want this (as in the previous case), but they also know that I have this intention and takes this intention as into their will as determining it. They want to satisfy my will, and I, on the other hand, presuppose and know that they want this; this is so in the current act and in the habitual decisions of the will. These are, therefore, the specifically social acts that presuppose mutual understanding of the kind that brings an ego into contact with an ego.

Husserliana 14, Appendix 24: Empathy and Sympathy. Sympathy and the performance of unanimous acts with the empathetic, memorial, and so on. <Empathy and putting oneself into someone else's place in phantasy, in art.>[3]

The Humean theory of sympathy is, of course, false. One must see that sympathy as co-feeling the empathized feeling (current compassion, shared joy) is essentially related to the co-feeling in the reproductive act of consciousness. I can also "sympathize" with my own past.

It could be said: normal memory is itself something like sympathy; the memory, "it was so," and so on, is a reproductively modified perception. Here we have not only something mere phantasy modification, a mere reproduction of the doxa, but rather, at the same time, a *present* doxa. The reproductively modified doxa is not merely one to be "relived" again; this is shown in the possibly illusory memory. The memory is deceptive. I believed and still believe, or now I do not believe. In the normal memory lies the *co-belief*. One could say, this is the sympathy of believing. Likewise for every position taking of a pleasing act, as well as every desiring, wanting, and past action. It is still pleasing now, I go along with the earlier desiring and having-decided, I act along with it or not.

In the same way, I sympathize with others, just as empathy is a modification of memory.

First: if I reflect now, in the midst of my present current life and reflectively examine it, considering whether I value rightly, judge rightly, act or am about to act rightly, and similarly if I reflect with regard to the past and ask whether it was right or not, [186] then I depend on the retrieval of counter-motives that could possibly cross the others out.

With regard to the other, I try this approach: empathy is something like remembering-oneself-into-the-other, and, hence, a sympathizing would take place just like in memory in the usual sense. But is it really the same situation for both of

[3] <*Husserliana 14*, 185–191.> <From around 1920, perhaps summer semester 1920.>

these? In memory, the present position taking is a *continuation*, an *extension* of my own old position taking, which has remained in me habitually; in empathy, there can be no talk of this. Now, one would like to say that, in empathy, there is a certain mediation, a presentification by indication.

But how does sympathy, the participation or non-participation of the acts, come about? Through "putting oneself in the other's place," which, however, belongs to every act of empathy; it is itself putting oneself in the other's place. But there is not first an other for me, and then I put myself in their place. What is here referred to is only a deeper living-into them, feeling-into them. The apperception "alien human being" has its basis in my self-perception, it is an analogical apperception, and since I am always there, I am also always in current coincidence of similarity with the other, who is only there for me by analogization. Admittedly, greater accuracy is necessary here. The other is constituted as a reproductive variation of me, while I am myself at the same time there and involved. To understand the other, to imagine them and to have them more or less clearly there is, at the same time, to present, to feel, to want in such and such a way.

Thus, it is similar to when I immerse *myself in phantasy*, as if I perceived, imagined, thought this and that and valued it so, and so on. The mere perceiving of phantasy cannot be crossed out. If I take it as my basis, a judgment, a feeling, a will "in the phantasy" can, however, be taken up with sympathy or antipathy; like in fairy tales, novellas, works of fine art. Here I am in the attitude of as-if.

How is the phenomenon to be interpreted? In a certain manner, I live in the other, but now I judge, evaluate, act "in a certain manner," in such and such a manner, but I, the one who is currently empathizing, I have my current motivation and possibly do not do with the other or also do with the other. When I phantasize myself into a phantasy situation, I know myself there as I, here I indeed have a memory as if. I—the current I—presentify myself in a presentified environing world, and discrepancies arise. I—the current I—have my real past, have my acquired peculiarities, my convictions, my habitual valuations and so on; and if I think myself into a phantasy situation, then there is a positing in it, perhaps with a real primordial doxa, like if I take myself to England. In this case I have an "assumption," "thereupon" I do so and so, I with the habitual make-up of my motivation. In this case everything is correct given this positing. [187] But I can also phantasize otherwise, then I have disharmony.

In these cases, free phantasy will thus necessarily run according to my habitual self or contrary to it. Of course, matters are much more complicated. The world into which I fictively place myself is a fictive world as well and presupposes my habituality. But I can live one-sidedly in a layer of the world in such a way that the phantasy violates certain experiences and experiential convictions of a theoretical kind; and I, the actual I, can become more or less sensitive to this; I do not go along with this, reacting inwardly to it with "nonsense," "that is impossible." It is like this in the layer of an imagined nature. But only then follows the layer of spiritual act, of valuation, of reasonable actions.

In the phantasy in which I playfully place myself into phantasy worlds or into phantasy situations and then act "dreamingly," in this "memory as if," there arises generally, in the letting-myself-go, the letting-it-come-to-me, in the phantasy, and in the dreamingly doing of all sorts of fitting things, no conflict in the intuitive realm. Accordingly, here I surely have no "sympathy," but <I do have,> in the contrary case, when I arbitrarily make up something against these subjective possibilities, a crossing-out, a corresponding contradiction. Another case, a distinguishable one, is dreaming, where dream images appear in passivity and I really value, act, and so on, in such and such a way, whereby I only afterwards feel the "unfitting."

Art: through the means of art, indications are created that draw me into a phantasy as something demanded, and while this is being completed and a presenting, valuing, acting as-if is being performed according to the demand, that which is or is not in accordance with the demand comes to my attention. And I—the I who coincides with the other I in a certain manner—*sympathize* by reproducing what is first indicated and symbolically grasped, intuitively, but <in such a way that> the demanded position taking, and thus the concretely required in the alien inner life, receives such an intuitive deepening that it really becomes one with the I, appears as emerging from it itself. Only then is it an intuitive and given *possibility* for me. That just means that only that which *is my own possibility* for me, the I who in empathy has the presentation of an alien I with whom, however, I am *eo ipso* in coincidence, creates a *concordant* coincidence—a co-positing, a co-judging, co-valuing, co-desiring, and so on; and, corresponding to this, a common environing world furnished with the same spiritual characteristics. In the contrary case, "I do not agree," is a conflict between how the other I takes a position, perceives, judges, values, and so on, and how I "would judge, and so on, in their place." *But I still do not yet have sympathy.*

I have shared joy with the other. [188] That does not mean that I would enjoy myself if I were the other or if I were to think of myself put in their place (and I would not, as is the opposite case, I would behave quite differently, mourn), but I am really joyful. In compassion, I really suffer. So, is it wrong to relate sympathy so readily to memory? In the memory of the past, I co-judge, co-value, and co-will and, as my actual I, really take a position in it. That seemed to be what they have in common. But in empathy (we always take fully intuitive cases, which, by the way, is probably not necessary), there is, similarly with hypothetically placing oneself into a possible situation, also a real position taking included in a certain manner, that is, as mine, of the empathetic subject. There is, namely, a *co-doing*, co-living in perceiving, thinking, feeling, deciding. It is, of course, not like when I hypothetically put myself into such a situation and then do it, as it were in the postscript (which, of course, is not a conclusion, an idea of "necessary consequence" is not present here!). But it is nevertheless in the essence of the empathy situation that, while I posit the other according to the indications as perceiving, feeling, and so on, in such and such a way, I myself, by performing the presentifying of each "I perceive," "I value," am precisely in coincidence with the other I, and have, as the current I, a common given *ground*.

I already have the current position of the other lived-body in the peculiar mediation that the position of my lived-body is constantly implicated in the position of the other's and is in peculiar "coincidence" with the alien lived-body. It is the same with my material environing world and that of the other, which is the same as mine, only seen from there, where my lived-body coincides, as it were, with the one there, and my ego, coinciding with the one belonging to the lived-body there, sees, from there and, as it were, with the alien organ of the body, this and that or what is "included" in it, which can be said evidently on the basis of the factual situation: He sees from there and has these and those aspects, I would have just the same from there. This gives the common "ground," namely, for the act-life of the other and for me. The passive environing world is one and the same, but the I is not the same. Nevertheless, I myself as ego coincide in empathy with the other ego, and, in their being-affected and being-active, I, who coincide with them, am, as it were, affected and active: I am it "really" as co-active if the acts of the other run as my own pre-figurations of motivation would make it possible, if they proceed just as I would really act on that *ground* and in the changed situation from there, opposite my current one, and under certain conditions that just this situation brings with it (he is attacked and I am not, and so on). I can say that the other person is only determined for me through what I can find in common. We must exclude knowledge, at least in the first examination. [189]

I do not have a hypothesis, like when I arbitrarily give myself a fictive ground. Similarly as in *art*, when I have a ground given via indication, on which I place myself. In art, I have an artistic demand (emanating from the work of art) on me that I accept, I have, thereby, a ground that I myself let be the case, which is from now on "there" for me, which I hold onto and which I further develop according to the demands (indications) of the work of art, and as soon as persons appear, I am in coincidence with these "accepted" ones, and now I agree with them or disagree with them: I have either unity, I as the actual I, "placed" on this ground, doing with, as the placed ego am not, of course, the current one. But here, there is an essential twist: I would behave this way and that way, I could behave similarly, I could think this way, and so on. On the other hand, I can approve of such behavior. In the contrary case: I experience, as I put in their position, the disagreement, the No, I can say as the present I: I would not do that. I can also consider and clarify, and only then, as it were, do, participate, or not participate. This, then, concerns phenomena of approving, and possibly of approving judgment, evaluation, volition, in higher acts of reason. *This does not result in sympathy.*

How is it, then, that I not only co-judge, co-suffer, co-value, co-will as placed in the other's position, but rather, as the current ego myself (in my position, on my ground) and as actually judging, feeling, wanting, and in the same sense as the other? What he judges, now I judge as well, I make judgments about their environing world, which is also mine, with which I participate. He suffers from what has befallen him. I suffer not only in the modified form that of suffering should I be in the place of the other, but I now suffer myself; I am sorry that this and that happens to the other. He is sorrowful because his mother died, because his fortune is lost, and so on. I am sorrowful because of it. But I am also sorrowful because he is sorrowful. What has primacy here? Is his mourning, the fact that he mourns, the object of my mourning, or is it primarily that what is sorrowful for him is sorrowful for me?

If I judge as he judges, that which is judged "in" the other passes over into my own judgment of the selfsame situation. This is an essential transposition. It lies in the situation of empathy, in the case of harmony, that, if I were in his position, I would judge in this way. But his situation is, insofar as the common environing world is the basis, one of those possible situations for me, which the sense of judgment "it is so and so" includes in all cases where I do not make an immediate perceptual judgment, but rather make a judgment on the basis of mediation as well as intuitive indices, in which it lies just so: If I were to go there, which I can do, I would be able to convince myself of it in a perceptual judgment. So, here at any time the transposition can occur, which makes one's own judgment about the same state of affairs that the other one has judged a judgment that I have empathetically participated in and "implicitly" taken part in. [190]

It cannot be quite the same in the case of willing. I cannot turn the other's decision into my own, however much I may participate in empathy. Indeed, what stands before his eyes as practically willed, the same can now stand before mine too, and, in a similar transposition, the empathetically given "this should be" and "I want to do it" can become one for me. But if he has taken hold of something, this generally excludes my own taking hold of it, and this gives rise to a peculiar discordance that has nothing to do with disapproval. Maybe, if he wants to push a boulder, for example, I can intervene, even though this already brings along a new complication. In his effort to provide a proof: there I can have a will attuned in the same way, there it does not interfere and the intervention on my part does not necessarily have the character of someone who helps him.

In *helping* there is: I want that his "will be done," I want his will to reach its goal in action, and of course my action belongs to it too, which, wanting to realize the same, gives a power for accomplishment that is connected with itself, whereby the accomplishment becomes a common one in which *each* will "is done."

Similar to the case of judging is the case of valuing something objectively valuable and valued. But there is also a value for me that is without value for the other, a value that is pleasant, pleasing to me, that is not for the other. What I enjoy and use up in pleasure is not enjoyable, pleasant, and so on, for any other. The possession that I have, and as long as I have it, cannot be another's, and thus, its value to me also cannot be current value for the other.

The mourning of my mother's death is not the same for the other, mourning of my mother's death, a mourning of the *same* content, as independently possible in each of us as the same judgment of the same state of affairs, for instance, even a mathematical one (at least an objective one), where I can indeed come to the same judgment by way of empathy but my judgment would be possible without empathy. If, on the ground of empathy for the other's mourning over the death of his mother, I now find it sad that she is dead, then this mourning of mine has the intentional and essential relation to the other's mourning, and I take part in it, which indicates, I am sad that they mourn and that they suffered the loss, and *therefore* I also mourn over the death of their mother.

When I empathetically grasp that a higher value, such as an important insight, an aesthetically beautiful thing, has become his own, then perhaps I can myself appropriate exactly the same theory, exactly the same subjective realization of the theory in the form of gaining an insight as well as an equally perfect "appropriation" of the ideal work of art, but each appropriation is a new one, [191] and so is every subjectively appropriated value as such. I can now value the fact that the other has such an appropriated possession, a spiritual one appropriated to them, I can rejoice in it, and mourn that they have lost it. Sympathy is a feeling and a value that proceeds in the same way. In sympathy I do not live in empathy, I do not think, feel, want, in the other and do not perform my approvals and disapprovals, namely my co-thinking, co-judging, and so on (possibly my Yes and No) only as my ego coinciding with him. But in "sympathy" the other ego and his act comportment is the subject of my valuing, my loving, and so on. Here is the authentic domain of the concepts of love and hate.

Chapter 6
Husserliana Materialien 9: <*A priori* doctrine of value and ethics.>

Edmund Husserl

We now add an important continuation of our examination by taking up the earlier distinction of cultural formations into those that, in intersubjective experience, prove to be factual entities, formations possibly recognized by most or all as valuable, and into those that can be shown in valuative reason to be true values. I remind you of our examples of the Renaissance sciences of alchemy or astrology, [119] which arose as spiritual constructions and were, in fact, generally held to be true sciences, though they turn out to be pseudo-sciences when put to reason.[1]

This leads us to the question of the principles and general laws under which stand all operations of reason in valuing, or, at least, all genuine, true values as such stand. Insofar as this genuineness itself emerges in a valuing, and a particularly distinguished valuing, which, for that very reason, is called "rational." Also, insofar as

[1] Crossed *out:* The question of the distinction between a natural scientific view of the world from a teleological one was, for us, the starting point for a series of independent considerations. We will have to continue them for a while before we can make use of the insights we have gained concerning the necessary demarcations in the realm of possible knowledge and topics of knowledge in order to understand why antiquity did not push through to a natural science in our sense and why the teleological view of the world is the essentially predominant one in it, and vice versa, why the modern view of the world with natural science developed a specifically natural scientific view of the world, but for its part is inclined to miss the peculiar sense and the necessity of a teleological view of the world.

<Edmund Husserl, *Husserliana Materialienband 9. Einleitung in die Philosophie. Vorlesungen 1916–1919.* ed. Hanne Jacobs. (Dordrecht: Springer, 2012), 118–169.>

E. Husserl
Freiburg im Breisgau, Germany

S. Heinämaa, A. J. Steinbock (eds.), *Values of Love and Ethical Reflection*, Husserlian Legacies: Themes for the 21st Century,
https://doi.org/10.1007/978-3-031-68698-6_6

this emergence is itself a spiritual achievement as a striving for and attainment of genuineness, we are thoroughly in the spiritual sphere, and the disciplines of reason belonging here are themselves to be regarded as sciences of spirit. Thus, in the most general framework of these disciplines, we come across the highest normative disciplines, we come across the *a priori* science of values and, therein, aesthetics, and furthermore, across the *a priori* science of practical reason, ethics. We will also have to say a word about logic in this framework and assign it a place here.

Let us begin with valuing, a general title for any kind of pleasing and unpleasing, being joyful about something or being annoyed by it, and whatever else the designations with various differentiations may be. We now point out a radical difference. When we value, that which is valued can be real or unreal, normally perceived or illusory, a free fiction. Consciously, the reality can be posited as reality, that which is unreal can be conscious as unreal. Sometimes we value things of the experienced environing world, sometimes a centaur-landscape that makes a picture intuitive for us while, at the same time, we take the fiction as fiction, sometimes we give ourselves over to the play of our own phantasy, for example, to the musical phantasy and take pleasure in the melody that is heard inwardly, as it were, in the consciousness of its unreality. [120]

Now, there is a valuing for which this given belief or conviction of the being real, the not being real, or even the being probable, and so on, of the valued object, is essentially motivating, fundamental for the position taking of the valuing. And there is, on the other hand, a valuing where this conviction of the real being of that which is valued (thus also every logical position-taking with regard to it) remains completely irrelevant for the valuing. In the first case, the valuing is directed to the object itself and how it is; in the second case, it is directed only to the appearing object as such, to the how of the manner of appearance. We delight in our comfortable apartment or in the beautiful days of winter. What is valued here is the relevant object-base in its object-based qualities. The valuation depends on the true being of these qualities. We take joy in a person, joy that he is just the way he is, that he has this intelligence, this heart, these musical abilities and skills. Then the valuation is based on the conviction of the reality of the person and the reality of these qualities. As soon as the conviction of the real possession of these qualities is lost, the joy is soon gone as well. It is the same with the displeasing, with the hatred of a human being whom we perceive as having qualities that repel us. Hatred loses its ground when we realize that we have been mistaken, that we have falsely attributed such qualities to him.

But we can value a human being as "beautiful," for example, a person as physically beautiful, in a wholly different manner. Obviously, it depends here exclusively on the manner of appearance: the beauty is only given to us, and the pleasing-beauty is only a really original and unfolded one, if the person stands in a certain position to us, if they show us certain aspects and not others. They may be recognizable as the same in infinitely many other manners of givenness, but their beauty belongs only to the distinguished modes of presentation. Likewise, a mountain is called beautiful only in relation to the view it shows us from a determinate viewpoint.[2] [121]

The distinguished view, however, is not a real property of the object itself, but a property of the perceived object as such in the medium of the manner of givenness. Whether the human person or the beautiful mountain turns out to be an illusion, the pleasing still remains as pleasing in the beautiful, so long as the intuition shows the object in the same aspect. Thus, pure beauty does not belong to the reality. If we understand "interest" and "being interested" in a concise sense, that is, in the sense of being interested in reality, as Kant does, then being well-pleased in the beautiful would be a "disinterest." If we designate the real mountain or the real person as beautiful, then this is a subjectively objective manner of designation: he is beautiful insofar as he, not only is as he is, but rather, insofar as he has certain manners of givenness intuited by a subject under certain aspects that found a pure pleasing-beauty. We characterize this pleasing-as-beauty in the broadest sense as aesthetic.[3]

This results in a great distinction of values into values of existence (Good-values) and into aesthetic values or beauty-values, and thus also in a distinction of the doctrines of values into a doctrine of goods and into an aesthetics. If, in pure aesthetics, objects are generally considered for the sake of their beauty, without regard to the question of the reality of the objects and their properties of reality, then it is now to be noted that, where a real object is pleasing as beautiful and is assessed as beautiful, the reality of this beauty can also, at the same time, arouse a pleasing, and it can do so precisely with respect to that it is a beauty. Said more clearly: we enjoy it, it is a value to us that an object is real, to which belong modes of givenness in relation to the apprehending subjectivity, subjective modes of givenness that found a pleasing-as-beautiful. [122] This, however, does not result in a mixed region, rather, it is manifest in an *a priori* law that, with every aesthetic beauty, a real object simultaneously accrues the good-value. As a real object, a thing has, besides other goods

---

[2] Crossed *out*: Let us start from the human being and their factual environing world. It faces them as existing, be it as mere nature, be it already equipped with cultural characteristics. It stirs their heart, they valuate it, and this further determines his praxis, in which, in the manner discussed earlier, the environing world takes on new and thereby also lasting characteristics of a spiritual kind. The environing world faces them in the form of experience as being there, as being certain, or, in the particular, also as being presumable, probable, and so on. In phantasy, the human being can also have the formations of the phantasy in their mind, and they can also take a position with regard to these formations in the way that they posit the phantasized as possibility, that is, as possible real existence, and then, as a rule, posits it in relation to a certain actuality. Thus, for example, when the human being considers how the possible, which is considered valuable and which appears in desire as what should be, can be realized willfully, can be realized in action. In general, we have here characterized real and possible environing world as a real sphere of being that is valuable according to reality and modalities of reality and is therein determinative, by virtue of the values attributed to it, for an action or activity. On the other hand, there is also a pleasing and displeasing, a positive and negative valuing, of mere phantasy formations as phantasy formations, not of phantasizing, but of the phantasy object, of the phantasy process, of phantasy worlds, namely, as phantasized in these and those modes of presentation, aspects, resonances of feeling, and the like.

[3] *Marginal note*: "Manner of givenness," this should be taken very broadly here: every manner of presentation, and so on.

also this good that it is appropriate, seen under certain conditions, to show itself as suitably beautiful. Thus, now we value its reality with and in a manner such that the originally pure beauty-valuation grounds the good-valuation: the object is not only beautiful, it is also good on account of its beauty. The evident connection that exists between pure beauties and goods that arise from the presentation of beauties makes it comprehensible that aesthetics, forfeiting its purity, also deals with beautiful objects and works of art as products. (It should be noted, however, that the joy in the creation of the work must also become aesthetically meaningful; that one is aware of that which is beautiful as the product of an admired artist and his creation, that is itself a mode of givenness of the beautiful that can have an aesthetic effect.) Let us now content ourselves with the mere prospect of a scientific and pure aesthetics as a science of pure aesthetic values, the closer consideration of which, in *a priori* and empirical terms, would take us too far afield here.[4]

We can give expression to the radical distinction of values made in the last lecture in the following philosophically valuable way: by "regions" we understand the highest material genera, whereby material was not any merely formal-logical determination. Nature, then, is a region, but value is a region too, and the classification of values into beauty-values and good-values was a fundamental distinction of the regions into two main genera. [123] I would like to add a second general distinction that permeates the realm of values in general, but which is not a proper generic distinction, namely, the distinction between self-values and values for the sake of another. For example: a person who is valuable through what he is in himself, and a dear memento of a person. Each of these main categories can now be treated separately, and, first of all, within the sphere of beauty, we can again ask about the basic genera and about their essential subspecies. I would now like to say a few words about this.

First, in the aesthetic sphere: one such basic genus of the beautiful is purely sensible beauty, which Plato already saw in its purity in the *Philebus*, for example, the purely sensible beauty of tones and tone formations. But not music in our sense, with which the spiritual, which expresses itself in it, is always connected; as the titles such as "Eroica," "Pastorale" already indicate, not to mention the expression of musical forms of language such as "songs," "operas." Likewise, a purely sensible

---

[4] Crossed *out*: Let us consider instead the realm of good-values, specifically in universal generality. Every will aims at a goal of the will, and this is obviously a good-value or, better yet, a good-value intended by the one willing. It must also be intended in the conviction of the one willing and acting as an achievable one for them. Thus, the good in general and the practical good are distinct. The latter concept is a narrower one, but the narrowness is related to the additional thought of a subject of the will and the respective domain of its practical possibilities. Let us now attempt to sketch out the sciences here. First of all, a science of goods, a theory of goods. Here it is a matter of the various kinds and genera of goods, but also of their comparative value, of the establishment of rankings of preference; furthermore, insofar as we consider the human being as a subject of will, a practical doctrine of goods. Of course, every human has their practical possibilities, their range of practically accessible goods, which even changes from time to time. But it goes without saying that, nevertheless, general laws for the determination of practical goods and, above all, for the selection of the practically best can nevertheless be established.

beauty of colors and color constellations with the addition of the sensible-spatial figures.[5]

We contrast purely sensible beauty with purely spiritual beauty. Among them belong, for example, the beauty of theories, such as, for example, the "elegance" of certain mathematical proofs and theories as distinguished from others that are just as good and prove the same thing but are inelegant. In this sense, we admire whole sciences, such as the peculiar beauty of Euclidean geometry, the peculiar beauty of ancient philosophy, and we do so, not for the sake of their truthfulness, but regardless of full knowledge of their falsity. Furthermore, every cultural sphere has its forms of beauties and not only scientific culture. We can also, in the sphere of the beautiful art itself, find a second valuation of beauty, for example, in forms of beauty in the whole development of art, a valuation of beauty which can take place in all parallel developments.

Just as cultural formations of an extra-personal kind can be valued as beautiful as a whole or in certain sides, so, too, persons, spiritual individualities, [124] (for example: we can value the beauty of their manners of acting). Likewise, personal associations, figurations of personal community, can be valued as beautiful. The beauty of a political development, of a history, the beauty of a certain state, the beauty of a religion, of a church, even if one is a non-believer. Spirituality of any kind gains its communicability and objectivity, so to speak, through a sensing lived body, and so spiritual beauty actually never confronts us purely, everywhere the beauty-value is a value from the connection. The spiritual finds its expression in the sensible, which has its own beauties, but which let their brilliance shine over into the beauty of the expressed, and thus result in a new, higher form of beauty. Of course, problems arise, not only of the classification and assignment of the laws according to which new values could be constructed *a priori* from beauty-values of one group or another, but also problems of the relative order of rank of the value types among themselves.

If we now proceed to the genus of good-values, the corresponding problems of a universal, not only empirical, but, above all, *a priori* doctrine of goods arise here. Thus, again, problems related to the classification of goods and to the laws of derivation, so to speak, to the possibly *a priori* laws of operation according to which goods arise again from goods. Finally, also the problems of the order of rank in the realm of goods, in particular, of the principal order of rank of the goods of different genera. To name a few headings here: thus, there is the class of aesthetic goods, encompassing all that is good on account of any sort of beauty; the logical goods: individual truths, conclusions, proofs, theories, whole sciences are valuated as values of being ("being" is here the logically correct, true); the goods of persons and of the personal community: character traits of persons or persons in their character traits can be valued as good, but so can particular activities of persons, such as acts

[5] *Marginal note*: It is enough for us, for reasons that can indeed be easily understood, to consider self-values. It should not be overlooked that when we speak of values as such, we mean either positive or negative values. (Strictly speaking, values are properly divided into three groups: positive values, negative values, and *adiaphora*, zero values. The zero is, as a value, indistinguishable.)

of love of neighbor, but also character traits of a people, of a nation-state, as they show themselves in a relatively constant manner in the course of their history, likewise a state in its constitution, in its establishment of particular norms, and so on.

The systematic study of a pure theory of goods, and, more exactly, of an *a priori* one (purified of everything accidental in experience) would have to investigate the laws of conjunction that exist between such groups, without which their systematic order cannot be obtained. For example: in the valuation of the person, we come across the valuation of their acts, [125] and we see that every valuable act gives value to the person, but also that the value of the acts depends on that at which the act is directed. If a judgment is directed toward a genuine truth, the judgment is a good-value for its sake. If a valuing is otherwise directed to a true value, then the valuing is *a priori* good if it values positively what is positively valuable and negatively what is negatively valuable. In the opposite case, however, it is bad. Likewise, the valuing-as-better of something truly better is good, the valuing-as-lesser of something truly better is bad. So, there are many kinds of *a priori* laws, according to which goods in the extra-personal sphere have, so to speak, their variants in the personal sphere.

Let us now make a further step from the doctrine of goods into ethics proper. Every will aims at a goal of willing. To the essence of a goal of willing belongs, *a priori*, that it is an intended good for the one willing; that is to say, that the goal is something presentable, but which the one willing lacks and so desires. Thus, it comes to mind as a non-being and as the end of a possible action that would realize it, that would transform it into a good that exists and fulfills the desiring- and willing-intention.[6] In accordance with this, the good *simpliciter* and the practical good are distinguished, namely, for a subject that is acting or is thought of as acting. Not every good can be *a priori* the goal of my willing; it must be possibly realizable by my action, according to my conviction, at least, I must consider the realizability as possible to some extent. Where I have the opposite conviction, I cannot set myself a goal, for example, to make $2 \times 2 = 5$, or to make the earth stand still.[7] [126]

---

[6] *Marginal note*: This is not sufficient.

[7] Crossed *out*: Of course, my conviction can also be mistaken, and, on the other hand, a supposed good that is practical for me need not truly be a practical good, even if I could achieve it and even if it were a good. If I realize a sensual pleasure, which in itself would not be bad at all, indeed would be good in and of itself, then it is a good. But if, for the sake of this sensual thing, I deny myself, say, a high good of love of neighbor or love of God, then in the sensible pleasantness there is achieved not a practical good, but something very evil. The leading sign of the practical good has reversed. In accordance with this, it can be said *a priori*: The realization of a +/– value of a good creates itself necessarily a +/– practical good; but how the leading signs coincide or are reversed depends on particular laws, such as, for example: the realization of something better is or can be a positive practical good; but certainly the realization of something worse, though good, is bad. Likewise, it is clear that the realization of something bad need not itself be a practically bad thing but is so only if the bad does not have a superior good consequence, and so on. It then always follows *a priori* that for every agent in every practical case only one practical good as such exists, namely, the best within the totality of practical possibilities concerning doing and refraining from doing; for refraining from doing too must be taken into account and can possibly be better than doing, for instance, where in positive doing only evil could be done. Once the most general laws

Obviously,[8] an *a priori* lawful regularity expresses itself in these self-evident facts. It is indeed evident that it would be a practical counter-sense to strive for something that, according to one's own conviction, is not practically achievable (capable of being done). One can also express this with the words: no one can strive for what he considers as unachievable. But then, one must not misunderstand this as a psychological law of nature. It does not mean that no one can factually strive for what is considered unachievable. For we know that such things occur, and unfortunately not infrequently. Every psychiatrist is familiar with this.[9] It is to be noted here that we call them insane, and especially those who are practically irrational, because they violate the immediately evident law of reason, either by not seeing it or by not letting themselves be practically motivated by it. The parallel case is that of the logically insane, who do not let themselves be inhibited in their logical convictions by immediately evident logical counter-sense.[10]

This *a priori* lawful regularity that determines the distinction between the good and the practical good is not, however, the only one of "practical reason," under the title of which we think of specifically ethical legislation. The one who is hopelessly stupid or mad, in practical terms, is not, for that reason, ethically irrational, that is, ethically bad.[11] What determines the concept of the ethical? Which peculiar *a priori* laws appear here? That this concept is a concept of value, and that the valuation governing it is not an aesthetic valuation but rather a valuation of being, is clear from the outset. It is similarly clear that what we thereby value are exclusively persons with respect to their decisions of will, [127] and specifically the goals of their will, both actual and potential. With respect to the latter, potential one, we assess habitual qualities of a person, familiarity with which shows how the person would practically decide under the respective circumstances imagined.

If we examine a person's possible goals of will, we distinguish two things. First, the question can be about the real value or the disvalue of the goal of the will, regardless of whether it is a goal of the will or not and regardless of whether it is valued by the willing person or not![12] Thus, scientific striving is not only directed to a supposed but to a real good, if the science or theory in question is a genuine and true one. But the good-value of this theory is a value in itself, regardless of whether

---

of this sort, independent of the possible, materially determined practical goods and bads have been clarified, then everything else obviously depends on the systematic doctrine of goods, namely, on a material doctrine of goods. In contrast to the impractical doctrine of goods, a practical one is needed insofar as it must be considered which definite kinds and types of practically realizable goods or evils there are at all, which enter as practical suggestions into the practical calculation, into the determination of the highest practical good, which is the only ought.

[8] *Marginal note*: Factually bad, whole rest of this lecture.

[9] *Marginal note*: This is not the way to go.

[10] *Marginal note*: Does this mean that they will forget the insight and it will not become a practically effective conviction for them?

[11] *Marginal note*: Quite inadequate; from [page <129>] on is retroactively patched up.

[12] *Marginal Note*: But how do we assess the volitional goals for which persons decide? How do we evaluate them and in accordance with how the persons in question evaluate them themselves? And this is the starting point of the considerations.

someone strives for it as a goal of the will or not and regardless of whether someone knows about this value at all or not. Second, however, goals of the will can also be valuated as goals of the will.[13] So, goals of the will can, as such, be positively- or negatively-valued. Here we are faced with the question of ethical value, of the ethically good and the ethically bad. Here, we encounter a double sense of the discussion of a practical good. On the one hand, it can mean a good which, intended as practically realizable, is a possible and possibly actual goal of the will; on the other hand, it can mean a goal of the will that is not only directed to a good and thereby to an achievable good, but that, moreover, has as a goal of the will its right, its good value. We can also express it like this: that something is a good does not yet mean that it is an achievable good for me, but it does say that it is, for me, a supposed and even an insighted good. And further: that something is an achievable good for me does not yet mean that it is an ought for me. Correlatively expressed: It does not yet mean that my will directed towards it is a right, ethical will.[14] [128]

Now, to make this clear: even if the goal of my will were an achievable and a good one, it could still be bad as the goal of my will, something that I ethically should not do. For example: perhaps one may seriously say that every sensible pleasure is a good in and for itself, say, a feast for the eyes or the stomach. If I realize such things, then I realize a good. But not all goods are on the same level in the scale of values. Thus, besides the sensible good in my practical realm, higher goods could be achievable. If I had the choice between the merely sensible good and the higher good of a love of neighbor or love of God, and I chose the sensible one, then this, despite its sensible goodness, would by no means be an ought, that is, not an ethical good; rather, I would have chosen an ethically bad one. In this context, the practical good reverses its sign; as a goal of the will, the intrinsically positive-value becomes a negative-value. *A priori* the law indeed obtains that the striving for an achievable +/– good necessarily creates or would create a +/– ethical good; but whether the signs on both sides overarchingly concord or not depends on special laws.

The following can be pointed out here: freedom of choice belongs to the essence of the willing I. It can be seen a *priori* that an awake I, an I that lives in performing acts, has a field of practical possibilities in every moment in its waking life. "I can" do this and that and all sorts of things, and can do so in such a manner that I am conscious of some things in the character of "what I can do." Further: it is evident to me that where I do something, I could also abstain, and where I abstain, I could do; that, preferring one practical possibility, I could also have preferred another one

[13] *Inserted page*: Ethical valuation, evaluation of persons, their volitional sentiment, volitional goals according to their motives, evaluation of their supposed values, and so on. Ethical ought: double sense: 1) the ethically right, the right ethical value. 2) What I should do, what someone should do, what is demanded of them, what can be expected of them, demanded of them; what I can demand of myself and what I, acknowledging the demand, fulfill and can fulfill, itself again in a will, in an action that can itself again be ethically evaluated. In an ethical valuation there is no question of an imperative. Concept of the imperative! The absolute ethical value—the categorical imperative.

[14] *Marginal note*: All of this is not sufficient.

of the practical possibilities; likewise, I could, surveying in advance my field of practical possibilities, decide in favor of any of the practical possibilities included in it. (If we speak of the practical realm, we do not mean the totality of what the agent could realize in objective reality, but what is before their eyes inwardly, consciously, as that which they are able to do.) It is a completely different law, which in no way disturbs this freedom of the "I can," that each of the decisions that take place has its motives, its "determining reasons" lying in the realm of consciousness of the "I." That I, among all that I can now do, do just this one thing is not decided from outside as in a lottery; rather, I choose just this because something in my consciousness speaks for it; and it speaks to me, it determines me to want just it, to decide for it. [129] For instance, I am tempted by a material advantage or I am determined by the superior intrinsic value of the thing. But that does not change the fact that I can do one as well as the other.

The ethical valuation refers to the will in the realm of its freedom of choice, this freedom that is nothing mystical at all but a perfect evidence. This freedom is evident for everyone who makes clear to themselves the sense belonging to the consciousness of will. Presupposing this, it is an *a priori* insight that the ethical good[15] or that which is demanded of any particular subject depends at every point in time on the entire practical realm as the realm of its conscious free choice and depends, more specifically, on the sum of the positive and negative presumed good values decided therein.[16] One sees at once that the best of this realm of the goods achievable for the respective chooser will determine itself as the ethical good, as the ought.[17] The exact determination and derivation of this principle here requires the formulation of a number of immediate axioms of ethical reason. One of the fundamental laws is that of absorption: if there are +/– goods to choose from in the realm of choice, which could be realized or not realized (done or abstained from) in freedom, according to the conviction of the agent, then of every pair of practically incompatible goods considered, of which only one could be realized, the lesser good is to be rejected *a limine*; it must, so to speak, be crossed out from the outset, it is to be abstained from. The will to the lesser good is not an ethically lesser will, but an ethically bad one. The better absorbs the lesser good in the choice; or, as the saying goes, the better is the enemy of the good. For example, no matter how strong the sensible inclination may be, the sensible good must not be chosen where, at the same time, a higher spiritual good can be chosen. Also, where relatively high goods can be chosen, the same absorption comes into question. On the other hand, a negative value can remain in the choice as a positive item and choosing it can possibly be demanded, namely as a means for an achievable higher good. All axioms belonging here must of course be expressed in a strict and complete conceptual

[15] *Marginal note:* Not "or," but rather as a further consequence!

[16] *Marginal note*: Supposed good value!

[17] *Marginal note*: 1) Ought in the sense of ethical value. 2) In the sense of what is to be to be demanded.

formulation. [130] They must take into account only the essential moments that are in question of the state of affairs of the will as such.[18]

Thus, for example, we have not yet taken into account the difference between means and ends, whereby we evidently come back to ultimate ends, that is, to ends which are no longer means for other ends. It is then an *a priori* law of practical reason that, under the hypothesis that an end is a good one, that is, an ethically demanded goal of the will such that every means appropriate for it enters the circle of ethical deliberation, that the hypothesis has, so to speak, a weight on the ethical scales. If the deliberation of final ends is still left in hypothetical indecision, then already in and of itself the realm of means relevant for these ends can be examined according to their relative weights and *a priori* axioms can be formulated for them: there are axioms of practical consequence, such as this, that a means that would bring about a final end with great probability or even with certainty is rationally preferred; the "better" means absorbs the inferior one, and the best means is the only one, as a consequence of the relevant valuation of ends, that is an ought, that ought to be if precisely this final end ought to be.[19]

All general axioms of practical reason are to be sought out. To be sought are both the laws of practical consequence, which concern only so-called "hypothetical imperatives," as well as laws of reason in the choice of ultimate ends of the will, not hypothetical, but categorical imperatives, if a multiplicity is even thinkable here.[20] In the latter respect, the question of the role of degrees of certainty in the selection of the "best" final ends must be answered, since also for final ends there may be a greater or lesser probability of achievement consciously there.

In other respects, it is also necessary to pay attention to what the relation of the whole respective range of achievable practical values to the moment of choice means. [131] The ethical judgment of a person—as we have formulated the concept of the ethical here and as concerns the goodness of their resolutions, or rather, of the goals of willing guiding their actions—requires that we consider them in a respective situation, that we empathetically put ourselves into their practical consciousness, that we reconstruct for ourselves the practical realm of what they are able to do in which they find themselves at that particular point in time, and then reweigh their decision. (Judging the person in their whole life and striving, therefore, demands carrying this very same process out for every point in time.) Here, however, it must be noted as an essential moment that no point in time is isolated and that the agent is, at every point in time, related *a priori* to an open future horizon; there is the practical sphere that determines and must determine choice and all the practical possibilities, from which an agent, if truly practically reasonable, can look forward, now into the future, and recognize as future achievabilities or posit them as

[18] *Note*: The whole lecture needs to be worked out anew. It was careless, hasty, that I did not speak of the difference of the objectively considered good and the good assumed by the agent, nothing of sentiment, of motives, and so on.

[19] *Marginal note*: Of course, if, on the side of the means, no other value differences come into question that are determinative.

[20] *Marginal note:* Here, there is no talk of the imperative yet. Cf. [<p.> 110 <= p. 134>].

presumptive ones. The best of the moment must be allowed to fall away, if its choice would make impossible a high good of the future. Of course, all good and bad consequences for the future must be taken into account in the question of what to choose now. It is clear that the future horizon, which is consciously present for every waking I, however different it may be in terms of its content, is constantly a practical horizon, demanding in it rationally positive deeds. An I can never be subject to the demand to do nothing at all, but, at the most, to a momentary holding itself back in order to be able to enter with an action, after waiting for a success. For we may posit it as an axiom that an individual in whose practical sphere there is no good at all is unthinkable. Also unthinkable is the demand to do something that would make any further future deeds impossible, for example, to commit suicide.

In any case, on the basis of such systematic and precisely formulable insights, one sees that, in the realm of practical reason (or of the ethical, in accordance with our previous formulation of this concept), there is not a multiplicity of practical imperatives for every agent and every moment of the agent's action but only one: the categorical imperative. There is for the agent, in each case and before reason, an *unum necessarium*; for the agent, a single thing is, in each case, the ought in each situation of the agent's life and for the whole of their waking and free life in general. If we construe the idea of a subject acting completely in accordance with practical reason, then the agent's whole life would be determined in ultimate necessity; but this does not at all mean that, on account of this, [132] the content of the future life could be calculated for each and every moment. For circumstances change in unforeseeable ways, and the presuppositions of achievable and of the value and the usefulness of the means, and so on, prove unreliable and in need of change in the course of experience.[21]

Let us consider more closely the sense of the categorical imperative and of all the *a priori* lawful regularities that bring about and support its determination.[22] The one thing that is demanded as such is to choose the positive good that is the best in the entire practical effective sphere extending, at the moment of the choice, into the endlessly open future. Valuated here are the goals of willing of any subject

[21] *Note*: I will probably have to give up the whole doctrine of the categorical imperative, or rather delimit it anew. 1) The *bonum* and *summum bonum*, considered from the side of the good-values (goods). The sphere of goods has for me a practically realizable part: my practically best "good"; 2) Is this already for me the ought? Could it not be doubted that the best practical good is what is for me the absolute ought? What is the meaning of the subjectivity of the will? Does it come into consideration only as objectified insofar as I evaluate it only according to the extra-subjective good that it creates? Problem of love. Cannot I have love for a value-region, and in such a manner that this love is not the same as valuing and delighting-in-the-value-that-one-possesses? A personal love, something specifically personal, which as pure love itself determines the value of the person. Now, one can distinguish values outside of me and in me and take this love itself into account as good. But is that enough, and so on? Love of the mother, love of the friend, love of neighbor, specific love of art in general, and so on. The beloved and the objective value. The beloved as, for this person, the subjectively pre-eminent one, and, on the other hand, objectively valuable in the objective assessment of the person.

[22] *Marginal note*: Not a categorical imperative, since there is no talk of an imperative.

whatsoever, so primarily and properly valuated here is the will,[23] more exactly, the decision of the will. For one must not take the word "will" in any loose sense, in the sense of a wish or an inclination or excitation of the will, but rather in precisely the sense of the determination of the "I do so!" that continues to work itself out in action.

It should be noted that the determination of the direction of the will, which the categorical imperative lawfully determines, must not be misunderstood as an objective one.[24] It should not be understood as if the will were good that chooses what is, in objective truth, the best for the agent among what is achievable and realizable. It is very well possible that someone prefers exactly that which is best among what is achievable out of blameworthy motives, for instance, because he expects a general advantage from that which, objectively considered, could be regarded as the truly best in the agent's practical realm. It will become clear how much it depends, not on the goal of the will, according to objective considerations of value, but rather on the inner motivation, the inner sentiment. [133] Already in the determination of the practical realm, it is not at all a question of the objective achievability of the possible goals of the will included in it, but only of the agent's conviction of the achievability of these. Likewise, regarding the goods and evils of the practical realm, it does not depend on whether they are true goods and true evils, but whether the agent values them as such, considers them to be such. In ethical valuation, the will of any person is valuated, not in an arbitrary sense, but in a quite definite one, namely, exclusively with reference to that which is conscious and so conscious as a motive, in front of him, to that which he takes as achievable, as good and bad, as better and best.

If we prefer what is best among what is achievable "according to the best of our knowledge and conscience," we have acted ethically. Our will is then and only then absolutely right. It is so even if it turns out afterwards that our convictions about achievability, about goods, the order of rank of goods, preferences were false. Of course, the subject can still be ethically blamed, not as the subject of the present choice, but rather as the subject of earlier volitional decisions. The subject could have, perhaps, in the past, better prepared judgments, could have better cultivated abilities of valuing well and of preferring rightly; in the past, this might have also been within the subject's practical realm, without being determined by it. Concretely, then, we must formulate the categorical imperative in this way: do at any time that which you can choose to the best of your knowledge and conscience as the best among what is achievable for you, decide for that. (For the concept of the best and the closer determination of the realm of practical ability that functions as the realm of choice, the axiom groups alluded to earlier have, of course, much to say.)

After all this, it is clear that the principle of the categorical imperative is nothing else than an *a priori* law of motivation, just as can be seen of every law of reason in the concise sense. It expresses an evident, absolutely valid lawfulness, how the motivation of a will must be such that the will can be an ethically good one, how what is willed can be an absolute ought. Indeed, not only that, for it also expresses

---

[23] *Marginal note*: What do primary and properly mean?

[24] *Marginal note*: Now then!

in the form of a law an unconditionally valid positive criterion of the ethically good and bad will.

It is easy to convince oneself that only the ethical valuation values the will according to a *summum bonum*, according to a practical best (according to a practically intrinsic value), while all other valuations of the will concern values which accrue to it for the sake of other given values. [134] The artistic will is thus certainly a good one, but only insofar as it is directed to the given value of the work of art and is the means of its realization. The ethical will, however, is a good-value in and of itself, even if what is willed might, objectively considered, turn out to be a disvalue. Let us now take a step forward. A rather peculiar step.[25]

To the realm of the will of each I belongs, *a priori*, the whole realm of the activity of its consciousness, including, therefore, its own willing too. Here, not only acts of particular willing, but rather of general willing come into question. I can, for example, make the decision to never, or from now on, never murder, steal, and do other good or evil actions. The I gives itself, in a will, a law for all further willing that falls under the lawful generality. And this will is not a mere momentary event, but a lasting sentiment, a lasting decision of will, which continues to have a conscious effect throughout the subsequent life of the will in the form of practical consistency. *Nota bene*: as long as a new, annulling decision of the will does not occur, and a new lasting decision of the will crosses out the earlier practical consistency and posits a new one; or, again, as long as the previous decision does not become powerless through further ruptures of the consistency on the part of differently directed, stronger motives or, through disuse, wherein the old resolution becomes powerless and inwardly, so to speak, fades away and dies. These fundamental *a priori* possibilities of will also come into consideration ethically, which means the general resolutions that fall into the practical realm must also be co-valuated and evaluated. [135]

What should interest us here is a curious application to the categorical imperative itself. So far, we have not properly had any right to speak of a categorical imperative. What we have worked out under that name up to now was the law of motivation and criterion for the ethically good will. It was not stated that the agent has

[25] *Note* 1) Judgment, naive belief—originally seeing belief. To every judgment essentially belongs, the possibility of being able to "question" the judgment, to make a project of judgment, an issue (problem) out of it and then to take up a position with regard to it. The judgment is, in the manner of grounding, insightfully motivated when the question (problem) is decided, motivated by insight, by going back to that which itself makes legitimate. Every judgment can take the form of a position taking on the problem, of being directed to the true itself. 2) Valuing, naive valuing - naive original value-ception. To establish a value problem at any time, to question the "value," to direct oneself to evaluative grounding, to consider valuable, decision of the question, valuing in motivation through comparing it to the "value itself" (value-ception). 3) Willing - naive willing. To every willing corresponds a possible question of will. Original willing, in which an ought is naively present, one is originally conscious of an ought. Every willing transferred into questions of will = to every practical goal corresponds the questionable goal, and the questionable is a "true goal," truly an ought, if the willing measures itself up to the original consciousness of the ought and proves its rightness each time; the goal is motivated (grounded) as right. But here come the questions of preference, relative ought and absolute one through the "higher" ought. The "lower" is the one that in the conflict of the "choice" has its ought-character crossed out, but only in this relativity.

formulated this law themselves, let alone that they must have formed a corresponding general will in themselves. If I do, to the best of my knowledge, the best of that which offers itself to me as achievable, then I do the ethically good. I do not need to have this in mind to formulate the generality of the lawful situation, and I do not need to see that there is also the practical possibility for me to willfully make it my possession to act in this way, that is, to make a decision of the will that has as its content to do from now on lawfully what is in each case best among what is achievable. But as soon as this thought comes to me and my sphere of choice is enriched by this willing law, it becomes an ethical demand and the highest ethical demand.[26]

The ethical criterion, the law of that motivation that characterizes a will as an ethical will, has the practical consequence for the one who insightfully sees this law that their will can be ethically good from then on only if they make the decision to always act according to the law. Thus, the law becomes the highest imperative. Act in all cases according to the principle to do the best, and so on. As soon as this imperative is presented to someone from the outside, as soon as it is communicated to the child by the one raising him and made comprehensible, it is *eo ipso* the ethical normative principle, the principle of real ethical praxis. At the same time, we see that a person who consciously decides in the sense of the categorical imperative to will to do good according to the law is ethically higher than a person who decides for the best but does not do the best as merely a consequence of a general willing of the law, to will to do good in general, a possibility that is conceivable.

Here,[27] we immediately take up the question of the ethical valuation of the person. According to what has been said thus far, what is ethically valuated in the first and most authentic sense is the will, and, strictly speaking, not the will as a momentary, fleeting experience of the one willing, but as a decisiveness of the will in the context of motivation that is at the same time enduring and continues to work forth in ever new modifications. [136] The person has ethical value as the subject of an ethical will, and gains, as the identical source-point of ever-new resolutions of will, a uniform ethical value according to the ideal which easily presents itself here, namely, the idea of a person who, in whatever phase of their entire life we consider them, is ethically perfect, that is, as one who exhibits that firm and conscious direction of will (=sentiment as a resolute sentiment) toward the thing that, in any given case, presents itself to him as the best, which the Stoics called διάθεσις <diáthesis: to arrange or put in order>. The ethical character then corresponds to them as the habitual peculiarity of the subject, as the original and, in the activity, firmly practiced disposition to the exclusively good will.[28] This absolute ethical perfection is thus a supra-empirical ideal. To develop out of unfreedom, out of the bondage of burdensome inclinations of sensibility and out of unethical resolutions and their motivating after-effects, ascending to "ethical freedom," to struggle through to the free resolution to want the good and the best for one's own sake and to provide this

[26] *Marginal Note*: How does the idea of a demand, a command, enter in here?

[27] *Marginal note*: <See the> next lecture [<= pp. 136–143>]. Thoughts good, presentation very poor.

[28] *Marginal note*: Admittedly, disposition is a dangerous word.

resolution with its effectiveness in the struggle with those opposing powers—that is the practical ethical task, the task of the constant approach to the ideal of perfection.

The scientific examinations that we have undertaken in the previous lessons belong to the realm of a uniformly self-contained discipline—ethics. It arises for us as the value science of the ethical will and of the intrinsic value, to be determined on the basis of the will, of a willing subject, a person. For the individual and in the individual case, the question of conscience or the ethical question constantly arises: "What should I do?" "What is, here and now, the ultimate ought and what is overall my true task in life?" Ethics as a science cannot deal with every determinate individual and every single case of the individual's praxis. It can, however, teach about the essence of a willing and acting person in general, about the essence of practical situations and realms of choice in general, and about the essence of that motivation in general that characterizes the person's will as ethical in generality and, first of all, even in the generality of supra-empirical ideality. (And there is even the curious thing here that the mere form of the motivation of the will positively determines the ethical value of the will and does not constitute a mere condition of the ethical value.) [137]

In the same *a priori* generality, one can and then must go further. Instead of speaking in mere formal generality of +/– goods that are available to the agent for the choosing, and of merely establishing laws that are formal laws of practical excellence and that culminate in the formal categorical imperative as a criterion, one can draw on the systematic doctrine of goods with the order of rank of supreme genera of goods treated in it, namely, pure genera. We can then think of *a priori* free subjects who have such goods in the realm of their practical sphere and theoretically consider what kind of life according to reason would have to result for them, which ideal systems of culture must arise there *idealiter* as demanded by reason.

Before we pursue this direction further, we distinguish the idea of a theoretical doctrine of values or doctrine of norms, as we have had it in mind so far, from that of a technique. This latter idea has its source in the fact that, as we have already recognized earlier, the ideal possibility of willing in general belongs to the essence of the ego. There belongs to the essence of the ego the ideal possibility of considering generally formulated practical possibilities and of making general resolutions, resolutions of lawful content, in which, therefore, the agent consciously prescribes for themselves a law of their further action, a practical law, and demands from themselves in all further action constant consequence, the remaining-practically-true to-themselves. This already holds in the formal sphere. The law of the supreme theoretical norm or of the highest criterion for an ethical will, which was emphasized in theoretical ethics, became the categorical imperative; and this indicates: ethics becomes the practical doctrine of norms, the technique of right action. The categorical imperative and all the consequences to be derived from it are taken up by the agent, who is guided by ethics, into their will; and that they should do this is itself included as a demand in the categorical imperative, *nota bene*, if ethics is practically there for them. This applies, further, if we take the pure genera of material goods that could become practically determinative for any conceivable subject of willing, as they are for us humans and, striving beyond formal legislation of will, establish material practical norms.

Here, first of all, the following is of great interest. There are, for example, as a genre of pure good-values, knowledge values, scientific values; another genus is that of aesthetic values and, through these, certain other good-values, like the artistic ones. And in this way, there are more of this kind. Each of them we can think *a priori* as practically realizable for rational beings. Hypothetically, we can now treat values of such a genus as supreme practical goals. [138] We take them on as ends and ask: How would a subject have to act, which kind of performances would it have to bring about in order to realize the best within the framework of such a genus? So, for example, within the generic idea of science: What kind of performance, of thinking activities, and what kind of theoretical formations in conclusions, proofs, theories would have to be demanded from any subject as a purely scientifically directed subject? Likewise, we could consider the idea of a most perfect creation of art and a most perfect art and do so in scientific generality. We would then seek general norms, laws, and value determinations for such formations. Regarding the evident possibility that these very laws could, in turn, be taken up into the will and thereby become practically determining for the actions of the scientist, the artist, and so on, in the sense of practical reason, techniques would then arise: the technique of science, the logical technique, as well as aesthetics as a technique or rather the technique of fine art, and so on.

All these techniques express imperatives, but only hypothetical imperatives, because, in them, science, art are treated like a final end, while, in truth, there is only one final end: the ethical will and the ethical deed arising through the ethical will. From this, it follows that ethics is not only one technique among others, but that all other techniques in general are subordinated to it and can only have the function of providing every agent with reasonable maxims, systems of helpful means, whether the ought is, in the agent's situation, scientific work or artistic figuration.

As much as all of this is correct, there is also a difference that is illuminating here. However many genera we could conceive of supposedly or arbitrarily hypothesized practical goods to be taken up and hypothetical imperatives related to them, that is how many techniques that we could execute in which the rules of practical consequence would be systematically derived, flowing from the respective hypothetical imperatives. Then we would have, for example, a technique of unrestricted individual egoism, in the case of the hypothesis: the greatest possible delight, the greatest possible power would be a good end, even a highest lifegoal. In the same way, we would have a politics, formed as a technique of unrestricted national egoism and national power, if the right goal for national life and action were characterized in just this way. So, in the former technique, technical, possibly scientific rules would be included for lying to, [139] robbing, and murdering one's fellow human beings, naturally, without being caught in the process and suffering harm. In other techniques, there would be rules for robbing other nations wherever one's own national advantage demands it, for enslaving them, for lying to them with idealistic phrases that beguile the heart, and possibly for murdering them bodily and psychically.

Of course, the difference of such hypothetical imperatives and techniques from the ones we previously used as examples stand out to us. For it was before a question of the highest genera of true goods, like aesthetic goods, goods of science, goods of individual and social personality, all of which are also names for practical goods, and which then become hypothetical imperatives only insofar as the formal practicality of the categorical imperative demands every agent to choose in their situation that which is in each case recognized as the best. Here, the following is a matter for further elaboration: it can obviously be demanded of someone that they choose science as a goal to be vocationally pursued for their whole life; or, for another, art; and, for still another, to provide helping and educating care, for instance, as a teacher; or, again, for a fourth, the parallel efficacy as a statesman or as a special functionary in the life of the state.[29]

According to the categorical imperative, another, second restricting relativity, and thereby, a certain, only hypothetical validity of a second kind, inheres in such vocational goals. At the given point in time of the choice, it can be required of a scientific researcher to put aside vocational work because, *hic et nunc*, loving care for others is the greater good; or, it can be required for an educator not to educate now but to take up arms for the fatherland, and so on. Nevertheless, the vocational goal remains valid for the person, whose categorical imperative includes it in the designated restriction as a permanent positive demand. Nonetheless, the situation here is essentially different from that of optional, arbitrarily constructed hypothetical imperatives and, accordingly, also from that of techniques to be developed by both sides.

Of course, we can even hypothetically consider a genus of that which is undoubtedly not good as if it were of positive goods. We can, further, materially construe a general lifegoal and consider it hypothetically as if it were a required one, that is, as if it corresponded to the formal rule of the categorical imperative, [140] though in truth it is not and, in the evaluation of the goal's content, it turns out to be an unconditional evil. However, corresponding techniques would not then be subordinated to ethics in the proper sense, namely, not as techniques for relative ethical goals, goals that can be demanded as ethical under certain restrictive circumstances; they would be, rather, subordinated to ethics only in the improper sense, namely, insofar as they, like all techniques in general, are to be evaluated according to the rightness and wrongness of their goals, and the principle execution of such an evaluation is precisely the task of ethics. Here, however, the judgment is decided negatively from the outset; it is a verdict that rejects in advance goals such as unbridled pleasure-seeking an unbridled seeking of national power, and thus shouts the eternal "No" against the corresponding, theoretically executable techniques.

On closer inspection, it is not only the hypothetical imperatives, the hypothetically assumed highest practical goals, that appear as valuated, but the techniques themselves that correspond to them. To build up a technique, to establish scientifically a system of rules according to which such a goal could be best realized, is

[29] *Marginal note*: Vocation. A general, practical theme that pervades life in terms of content.

indeed itself a practical goal, the value of which depends on the value of that goal determining it, that is, the technique itself. Egoism is an unconditionally bad life-goal, therefore, every technique of egoism is also a bad thing and even an ethically evil thing, provided that ethics proves to us that the goal is without value and must not in any way be included in our will.[30] This concerns individual as well as national and state egoism, in the same manner. An individual ethics of egoism is an unconditionally bad thing, which is why the hedonistic ethics of Aristippus and Epicurus (precisely as ethics of the greatest possible individual pleasure) has rightly and at all times been regarded as ethical skepticism or, rather, as negation of a true ethics, negativism.[31] In the same exact way, a politics, a technique of the state, that, as Machiavellian state ethics, pursues the idea of unrestricted state or national power is unconditionally bad. Thus, these techniques themselves as techniques are ethically bad. Whoever creates such negatively valuable cultural forms, collaborates in forming them, acts, if he has an eye for the badness of the goal, in an evil way, and he gains an eye for it through ethics. [141] On the other hand, techniques as such can be ethically good, such as logical techniques, aesthetic technique, and finally and in the highest sense the ethical technique itself.

It is itself obviously a task of the ethical technique to determine the ethically required supreme techniques. Closely connected with this is the ethical question of vocations. We find factually, corresponding to the supreme genera of practical goods, the genera of vocations. (After the deduction of the formal categorical imperative, there is to be reference to a classification of goods that can possibly become practical goods and there is to be sought rules as how to proceed in the competition of goods of these different classes. The situation would be very simple if it were in the general essence of an agent to be able to realize goods of all these classes, while, at the same time, it could be seen *a priori* that a single class would be preferred for the highest and practical good because, through it, the goods of the other classes would already be absorbed according to classes.)

If science, art, love of neighbor, love of God were separate genera of practical goods and practically incompatible, and if an order of rank existed that would distinguish one class, for instance, that of morals, the goods of love of neighbor, as the unconditionally higher and highest, then every scientific researcher and artist would *de facto* be an ethically blind or bad person, and then the corresponding logical and artistic techniques would themselves be bad as a consequence of this. But, of course, there can be no question of that. It must, therefore, be *a priori* considered what the conditions are here, and then which types of an ethical life and, at the highest, of a life illuminated by insight into value, and then which types of ethical personalities are *a priori* possible. If, in these types, goods of all the supreme genera can play their part, then obviously the relatively highest techniques assigned to them

[30] *Marginal note*: (All this interferes with the simple train of thought!)

[31] *Marginal note*: Too broad and divergent from the main idea.

themselves acquire the character of goods and of practical goods.[32] Then surely there will belong, among the types of agents, and indeed *a priori*, those of which are capable of taking into their will the generality of a scientific practical legislation and of letting themselves be insightfully and rationally guided by it and thus of ethically-practically raising themselves to a higher level, which is then *eo ipso* the one required according to the categorical imperative. [142]

Furthermore: certainly, among such types, there will also be the types of possible vocational human beings or agents of the form of scientific, artistic, pastoral, or political vocations, with which we are well acquainted. Their idea can be construed *a priori* under the guidance of the typology that is factually given in our factual human life, and which factually develops again and again in humanity. In all of our lives, goods of all classes appear as practically achievable, and, depending on the circumstances of the momentary situation, sometimes these, sometimes those are the ones to be preferred. God and ethics do not demand that we, lest we miss church or interrupt a heartfelt prayer, should let a murder happen or that we should let go of a scientific discovery that we have just gotten in our grasp, in order to search the streets and alleys for someone in need whom we could help.

However,[33] no matter how much, in every human life, the momentary goals of the will change and so too the spheres of goods to which they are assigned, still each of us has our "vocation," and our ethically required one. Not infrequently, this demand is a clear and completely unambiguous one in advance. Not a small number are told by an inner call of their "daimon" and a survey of the practical possibilities of their future life horizon confirms that they, all in all, do their best when they regularly devote a main part of their working time to scientific research and develop into a master at it. They must give the ethical shape of the relatively best to this basic stock of the goods they have achieved by linking it with achievements from other spheres of goods. In a noble sentiment, they renounce many beautiful and glorious things, precisely in the consciousness of forming their whole life in this way, on average, to the relative best. Another chooses the vocation of artist or actor, a third, that of preacher or educator. Of course, if need be, the artist could have made a scholar, the politician an actor, the actor a schoolman. But just for the sake of necessity. And in general, whoever does one just as well as the other will not be the best man for either of the two.

As soon as we value the typology of vocations from the point of view of practical reason, and thus value ethically, we necessarily come across the highest genera of value. It is according to the highest genera of value that the possible vocations themselves are typified from the most supreme points of view, then, further, to the highest techniques, [143] which are assigned to the highest genera of goods. These techniques are then themselves demanded by practical reason, endowed with the positive mark of value. As a further consequence, all techniques and their activities

[32] *Marginal note*: Read up to here. Cf. next lecture [<= pp. 143–151>], where has been repeated and then clearly continued.

[33] Two zeroes next to this paragraph in the margins.

and types of performances subordinated to the highest techniques; just as, for instance, all sciences and methodologies of these sciences are subordinated to the logical technique, down to their most specific, technical rules, as they are learned in seminars, laboratories, clinics. In the same way, a variety of special arts and crafts are subordinated to the technique of economic goods. And it is so everywhere.[34,35]

If here, parallel to the most supreme genera of goods, the highest techniques appear to be ethically ordered in the same manner, namely, from the point of view of vocation, then they all appear ethically only to the ethical technique. This is so since, from the ethical technique, all the regulation of reason proceeds, finally embracing all of them when it becomes more concrete and ever-more concretely developed. Ethics is, after all, the theoretical doctrine of norms and, furthermore, the technique for all rational action; practical reason in logical-scientific thinking, and hence reason in the form of all the sciences, is a branch of ethical reason, but no less so for practical reason in the form of artistic creation and in all spheres of any cultural achievement.

The thought that occupied us in the last lecture was the following: ethics, as the technique of good willing and action, has in the law of the categorical imperative a formal, but nevertheless absolute, necessary, and sufficient criterion for good action or for what is in each case the ought *simpliciter*[36]. The *a priori* doctrine of the most supreme genera of possible practical goods or non-goods, of the divisions and orders of rank within these genera and in the comparison of the genera with each other, then gives a *priori* principles of the application of the categorical imperative. This is enough to see that, as a consequence, all other techniques are under ethics, such that this is, as it were, the royal one among all techniques. All stand under its ruling. If the most supreme end of a technique is a bad one *a priori*, as, for example, the end of unrestricted seeking of delight, [144] then the technique is ethically repudiated as a bad one. If, on the other hand, the supreme end of a technique is a good one insofar as it refers to a genus of positive-value practical goods, such as science, fine art, valuable personality, state, then it is not yet said, without further ado, that the relevant techniques (the practical doctrine of science, aesthetics, educational art, state art) are already ethically approved. Their imperatives are still hypothetical imperatives, so long as it has not been shown that the supreme ends of these disciplines are demanded on ethical grounds, that is, are included in the formal demand of the categorical imperative, or so long as it has not been shown that the consideration of the determinate generic types of possible practical goods confers on the categorical imperative a form determined in content, which categorically demands from the agents a realization of goods of all such types, even if not in every single case.

This is how it is for us humans. For us, the formal demand of the law to do the best among all achievable goods to the best of our knowledge and conscience is not

---

[34] *Marginal note*: Did not read.

[35] A zero in the margin.

[36] *Marginal note*: recapitulation.

the only insight into our absolute ought. For us, as beings of a matured type of reason, it is evident that not only can goods of all classes (theoretical goods, aesthetic, personal goods, among them goods of love of neighbor, and so on) alternately appear in our practical spheres of action, but rather that we may not completely neglect goods of any sphere without violating the categorical imperative. Thus, it has a more concrete form for us, which, even if at first in some indeterminacy, enjoins us to promote goods of all genera, that is, to make ourselves skillful in advance for their cognition and realization.[37]

The categorical imperative does not demand for anyone, much less for everyone in general according to an *a priori* law, the exclusive realization of goods of a single sphere.[38] It is by no means as if a law of order of rank prevails between these classes of goods that would give an unconditional preference to those of one class or to a particular kind in all circumstances and for all persons, for instance, to the goods of neighborly love, of piety. If this were so, then practical goods of all these other classes, despite their goodness in themselves, would be once and for all negatively-valued. [145] But this is not the case.[39] In all of our lives, depending on the circumstances, goods of one class and goods of another class are the absolute ought. God and ethics do not demand that we, in order that we not miss church or interrupt a heartfelt prayer, should let a murder happen, or that we should let go of a scientific discovery that we have just gotten in our grasp, in order to search the streets and alleys for someone in need on whom one could practice neighborly love. For this reason alone, then, the techniques related to the categories of positive-value practical goods are endowed from the point of view of ethics with a positive sign; they represent, albeit in a conditional way, constituents of the highest practical good. For this reason, they are not only tolerated by ethics, but they are integrated as constituents of an ethics that is concretely carried out and does not merely remain in formal generality.

But we can still essentially strengthen the reasons by recalling the ethical vocational duty existing for the developed rational human in general. It should be left open here as to whether the necessity of a vocational life belongs *a priori* merely to the idea of the human or already even to the formal-general idea of the rational being in general. Not infrequently, the categorical imperative, in the form of a conscientious self-examination, points clearly and unambiguously to a determinate vocation and thus to a determinate class of valuable achievements and no other. For some, an interior call and urge of the heart points in advance to science, for others to art, to the vocation of the educator, preacher, and so on. This urge is confirmed or can be confirmed in a survey of one's own practical possibilities in the foreseeable future horizon of life, and it becomes completely clear to the ethically deliberating person that one would do one's best all in all if one would regularly devote a major

[37] *Marginal note*: Is it a duty for a Christ to go to concerts or collect paintings, and so on? Is not His task perhaps so great that all this "formation" disappears for Him?

[38] *Marginal note*: But what about the one whose vocation was to be redeemer of humanity?

[39] *Marginal note*: Difference between general and individual valuation.

part of one's time and best of one's labor to precisely this region of value, for example, that of scientific research, and developed into a master at it.

In advance, therefore, a mere basic stock of valuable achievements is distinguished, which is not in itself that which is ethically required, but which is to receive the shape of the highest practical good by linking it, first and foremost, with achievements from other spheres of goods as they are demanded by the respective circumstances of life. The one living in a vocation then renounces, in a noble sentiment, [146] many beautiful and glorious things, precisely in the awareness that he can shape his whole life into his relatively best life only in this form of vocational life.[40]

If we survey the vocational life of humanity and the manifold typology of vocations from the point of view of ethical reason, [147] then it is clear that they are all

---

[40] *Two Inserted Sheets*: It is clear that an ethics carried out according to the mere categorical imperative, as it has here, following Brentano, taken as a basis, is no ethics. I have completely fallen back into my old trains of thought; and yet, already in 1907, Geiger made the justified objection to me that it would be ridiculous to demand of a mother that she should first consider whether the support of her child was the best thing in her practical sphere. For the establishment of a norm of value as well as for the evidence of value on which the establishment of a norm of value is based, two things come into examination, and indeed first of all for singular material values: 1) the objective value, the value that everyone, every axiologically reasonable person who has the respective material content as a basis can grasp with feelings and value-ception, *originaliter*; 2) the same objective value as individual, subjective love value. In other words, what is meant there is this, that values relate to the valuating subject and its rational acts differently than objects relate to judging and generally objectifying acts in the logical sphere.

What is "value" at all, of course, is value in itself; every "reasonable person" can in feeling and value-ception grasp the value. But the same value can "mean" infinitely more for one than for the other. This meaning-more is to be considered first in relation to the so-called affective values. One not only values something, one has one's passion for it, one is then "in love" and the like in an "unreasonable" manner. But one will not be able to avoid recognizing a "genuine love," which is not only a value-ception of a beheld value, but a deciding for it from the innermost center of the ego, and indeed a lovingly deciding. It will then have to go further and say that what is loved in this way has a new value character originating from the respective ego, which is evidently added to the objective value insofar as it can be evidently found in it by the lover, but is added to it only for this ego. Everyone else, if he empathetically understands the lover as a lover, must recognize this subjective love value as the objectively true, but valid for this ego. Certain values, values of certain regions (spiritual values of each region), have not only objective value, they are also potential love values for possible and actual persons and have value also for their sake. Things then have values of transmission, like empirical values for the sake of being means for the realization of love values (like means of childcare). For practical reason, the "subjective" values now come into special consideration. The *daimon*, which leads to the true vocation, speaks through love. So, it is not merely objective goods and the objectively greatest good that matter, but rather everyone has their sphere of love and their "duties of love."

If I have made this attempt, the question is whether I do not have to go further. What would a life be without love? Love and value feeling must first be thoroughly examined. Should one decide lovingly for all values? But there, again, there needs to be distinctions within love. Can a child be an object of love for everyone in the same way? Every human is the object of a possible general love of humanity, which is ideally possible for and "demanded" of everyone, which is normative for everyone. But in contrast to the general human love there is a personal love that has its own right and, for this lover, confers on the beloved a legitimate individual value. Thus, the love of friends, the love of mothers. Also for higher personalities: the love for my people, which does not exclude the general love for every people.

subordinated, insofar as they are positively-valued at all (in the case of professional thieves, fences, pushers, I will, of course, hardly speak of vocation), to the *a priori* supreme genus of value, according to which the vocation themselves are typified under an *a priori* point of view. We then come across the most supreme techniques, which are assigned to the most supreme classes of goods. These techniques, like the techniques of science, those of fine art, are, from the point of view of vocational ethics, techniques demanded by ethical reason and integrated in its framework; they become components of a practical ethics. What has been said then extends to all specialized techniques subordinate to the supreme and principle ones.

This becomes clear by way of an example. Logical technique, which is under the supreme idea of knowledge and science in general, has under it all sciences and scientific methodologies; with them, then, all rules, even the most special ones, finally have their ethical sign, that of true, ultimately founded practical rationality (rules, as they are learned in seminars, laboratories, and so on). In the same way, the manifold of special useful arts and trades is subordinated to the general technique of economic goods after they have gained their ethical mark and their lawful limits from general ethics. And so it is everywhere.

In this way, then, we gain the insight that all techniques, that is, all systems of rules of practical reason, have their ultimate centering in the ethical technique from which they are all derived. This is so insofar as they have any real share in practical reason in general. If their most supreme purposes are ethically rejected, the reason of practical consistency may still rule in them; but, with the ethically rejected practical reasons, all practical consequences are also rejected; and, thereby, practical unreason prevails. The universality of the categorical imperative also gives universality to ethics, the universality of legislation in all possible practical spheres; thus, it itself is efficacious in all legitimate practical rules and techniques. Scientific reason, artistic reason, and so every kind of reason, insofar as it is conceived practically, turns out to be a specialization of ethical reason.

This is an almost self-evident truth, but nevertheless a truth of unprecedented scope. One laments the specialization in science, as well as the specialization in art, as a regrettable limitation. [148] But what is to be lamented to a much greater extent is the specialization of sentiment, of the fixed habitual direction of will, which is associated with the other specialization, but not only with it. One is a scholar, the other an artist, still another a doctor, a teacher or something else; but even if one is great in it, what is the use of this greatness and of the reason that is active in particular disciplines, if it is in its ultimate foundation unreasonable? The true greatness in art, science, and all practical spheres is, under all circumstances, ethical greatness. This, however, demands, on the one hand, that one, for example, pursues science for the sake of its particular value, that is, in pure love for the highest possible theoretical values of this sphere, that one not pursue it for the sake of fame or in the manner of a distinguished sport in order to rejoice in one's own and the others' superior skill. But it is also necessary to have an eye and a heart for the whole world of values in general and to regard oneself humbly as a servant to whom the advancement of a sphere of goods is vocationally entrusted, as someone who is destined to do his best ethically and thereby to advance, as far as possible, according to his own weak powers, the becoming of a world of God, a world of progressive fullness of values.

Admittedly, from there we come unexpectedly into metaphysics—for which we are not yet prepared. We have, however, come far enough to understand that ethics is neither a term for actions meant to enslave the agent nor to inhibit his ascension to the "overman." Ethics is, rather, a term for scientific, and, correctly executed, simple and fully evidenced norms and practical rules determined to give our practical freedom the form of the freedom of reason, in other words, determined to give our whole life and striving the form of the most beautiful and best as is conceivable, and so also the most satisfying life as is conceivable.

As a further consequence, it could be said: it is the task for the human being as rational being also to give to our environing world, which is, after all, at once with our active life one that is always being shaped anew, the form of a truly valuable world, a world of noble culture, which must have its unity and harmony in itself as a reflection of the unity and harmony of our personal life. The world must not, therefore, be stuffed-up with one-sided values or even with the products of a technology wasted on apparent-values, but it must satisfy the idea of realizing, and increasing from generation to generation, the best possible goods thanks to the action of an ethically minded humanity. This best possible world of culture would, *eo ipso*, by virtue of the practical relation of humans and humanity to themselves, [149] also contain within itself a best possible humanity, that is, a uniformly organized humanity, educating and developing itself to the will of a loving community.

Ethics demands all of this. But before we proceed from here to the general questions of worldview, some important additions are necessary. First of all, we want to abide by a sharply emphasized Kantian notion not to let the essential limits of the sciences flow into each other; he had especially in mind the separation of purely *a priori* and empirical sciences. In our previous considerations, we did not, as a matter of fact, keep the distinction entirely pure and clear; we must restore purity, since the highest philosophical interests are affected by it. Proceeding further, we must also emphasize the difference between formal and material ethics here. We have, indeed, repeatedly spoken of the formal law of the categorical imperative and characterized the axioms and laws belonging to it, such as the law of the absorption of lower goods by relatively higher ones, as formal. What was meant was understandable from the context, but now, it must also be fixed conceptually.

When we introduced the concept of the formal as opposed to that of the material in the theoretical logical sphere, within the correlation "cognizing thinking—truth—being," it concerned modes of cognition that belong to the essence of cognition in general, whatever it may be directed to, which thus remains indeterminately variable. It was accordingly a matter of objectivity in general and of all derived concepts that belong to the empty idea of an object or something in general, and likewise for truths. We thereby gained formal sciences, gathered together in the idea of a formal theoretical theory of science. (The formal logic of propositions, formal arithmetic and analysis belong to it.) In contrast to these are sciences, material sciences, which, whether as *a priori* or empirical sciences, take into account the relevant types of possible objects, dealing with, for example, natural things, with space, force, measure, and so on.

When we then went over into the world of culture, coming across value predicates and practical predicates, our way finally led to scientific disciplines of fundamental universality, the theme of which was the *a priori* of valuing in general and of possible values in general, the *a priori* of willing and possible willing achievements in general and oughtness in general. [150] From the standpoint of formal logic, the truths and sciences belonging here are materially determined. Value and disvalue are not formal-logical categories. (It does not belong to the essence of an object in general that it must be valuable or non-valuable in itself, or even that it must be an ethically good or bad thing.)[41] On the other hand, the concept of "value" and the concept of the ethical ought has a certain universality similar to that of the concept of being. If the latter leads us back to the consciousness of holding-as-being and especially of judging, value leads us back to the consciousness of the heart and of pleasing and displeasing with all its modes, and the concept of the ought back to willing.

If we go back to subjectivity, however, judging, valuing, willing are inseparably connected in it. No subject is thinkable without acts of all these classes, and so also unthinkable without relation to objects, to values, to goals. I do not here want to go into how these functions reciprocally relate to each other, how all objects, even if they are not valuable in themselves, can become significant in terms of value through relation to what is itself valuable through the transfer of value, how goals can necessarily be valuable, on the other hand, how values can become goals, how values and goals can be judged, and thereafter move into the sphere of being as objects. Enough, we have here three supreme functions, which demand a parallel consideration and may not be considered only under the guise of the object-positing function for the sake of something like that peculiar interrelation.[42] [151]

---

[41] *Marginal note*: This is unclear.

[42] Crossed *out*: We understand the necessity of transferring the concept from the formal to the value sphere, the axiological, as we also say, and the practical sphere. Just as, then, in general, the radical basic concepts of reason, like the concept of reason itself, form themselves parallel in a threefold way: theoretical (objectifying) reason, axiological, practical reason; being, value, ought; judgment decision and logical proposition, value decision and axiological proposition, volitional decision and will sentences. All reason expresses itself in the logical sphere by laws of logical form, in which value decisions, value sentences express themselves as value judgments and likewise will-decisions as will-sentences. We will have to characterize axioms of values and axioms of will as formal, if they keep within the supreme generality of value as such or, on the other hand, the ought as such, without determining the particular genus or kind of values. All our axioms established earlier were formal in this radical sense. What they say is valid for every valuating and willing subject as such and for values and aims as such, for goods as such, beauty as such; there is nothing materially prefigured that binds goods or beautiful things, like, for example, beautiful sounds or the beauty of visual appearances. Even when we say sensible beauty, we have a formal concept, because "sensible" does not mean something materially common, as, for example, color is a common essence that all determinate colors have in common, and likewise a tone for all determinate tones. Thus, the idea of a formal *a priori* theory of values (I used to call it axiology) and a formal *a priori* practice is determined as parallels of *a priori* formal logic (or philosophy of science).

It is the greatest philosophical matter to bring to light, in all spheres of reason, the pure principles and systematics, those of the principal structures, and to place everything ideal above them in the unity of a connected cosmos of ideality, and, likewise, to treat all empirical connections in the unity of the one matter of factual world and factual science. From the beginning and remaining forever a main motive of ultimate metaphysical problems is that this unified matter of factual world stands under ideas and ideal norms and, insofar as it corresponds to them, has its share in reason, and yet only a share, and is thus an irrational world. [152]

In the last lecture, we were occupied with restoring the idea of the formal *a priori* in the value sphere and practical sphere to its legitimacy and putting it in appropriate relation to the idea of the formal in the epistemic sphere. In the epistemic sphere, we had: 1) carried out the distinction between *a priori* and eidetic sciences and mere factual sciences *a posteriori*. On the latter side were sciences that, like the empirical sciences of nature and those of spirit, pertain to facts of experience; on the former side were sciences that, like mathematics, set apart what is founded in pure essence, what is generally pre-figured for objects as objects of pure ideas in unconditional necessity. In the sphere of knowledge, we had 2) distinguished sciences that belong to the formal idea of a something or object in general, and in connection with this, we had raised ourselves to the idea of a formal logic or philosophy of science that unfolds the *a priori* of the object in general, of truth and cognition in general in theoretical disciplines.

These disciplines are accompanied by material disciplines, which would deal with the genera and kinds of materially determined values and of practical goals. In ethics, which we wanted to more exactly carry out, we remain in the formal ethical sphere as long as we take as a basis the formally general idea of a reasonable volitional subject as an acting person in general. We may then use humans as examples, but we may not use anything that goes beyond the fact that the human is an acting subject in general.

Within the framework of this idea of a practical subject of reason in general, we can then pursue the free possibilities, the constructible types, insofar as they can be constructed with the means of purely formal concepts of value and will. As soon as we take the idea of humans as such as a basis, we have exceeded the formal sphere. We can then still preserve an *a priori* ethics of humans, a specifically human ethics, as long as we do not include anything factual, nothing that binds itself to contingencies of experience. In short, we form the pure eidos "human"; this involves typifying the human environing world in the manner of a pure *eidos*. If in carrying this postulate out it was shown that the pure idea "human" has a specific content that really goes beyond the formal idea of a rational being as such, then a specifically human theory of values and ethics would have to be separated from a formal one. But all this is a dream of the future, since nobody has made these postulates, which are highly important for a radical worldview, their own. The pure separation of formal and material *a priori* ethics correlatively requires a pure separation of an empirical and then, of course, human ethics. Historical ethics was such a human ethics, it wanted to be a technique, to give the human being as they factually are on this earth rules for how they could exercise practical reason under normal typical circumstances. Of course, some sentences in it were *a priori* and even formal-ethical; but however much the *a priori* has been noticed in certain philosophical directions, and however vividly it has been asserted, the systematic guiding ideas for the consistent separation of the *a priori* have been lacking, apart from the fact that mythical notions of the *a priori* spoiled everything from the outset. Other philosophical directions were and remain to this day blind for the proper legitimacy of the *a priori* in general and do not at all separate …*text breaks off.*

At this point, it should be shown that not only the difference between the *a priori* and the empirical, but also the difference between the formal and the material is transferable to the value sphere and the practical sphere for essential reasons. At first, this seems objectionable because value sciences and sciences of practical formations and of practical demands, like ethics, are special sciences; as such, they stand under the formal-universal idea "science in general," and so they cannot be likened to logic. Only this much was set: that there is an *a priori* of valuing and acting, thus that the sciences of spirit, whose peculiar basic concepts point us back to acts of valuing and acting, necessarily lead back to *a priori* cognition, which demands their systematic development in *a priori* value sciences and in *a priori* doctrines of the norms of praxis, and, most of all, in *a priori* ethics. From the point of view of logic, these sciences were no longer logically-formal: "value," "purpose," "means," "absolute ought," and the like are not formal-logical categories, not equivalent to concepts like "property," "relation," "whole" and "part," and the like.

On the other hand, these concepts like "value," "beauty," and "good," like "end" and "means" are not on the same line with material concepts like "color" and "sound," like "plant," "body," "earth," "physical" and "psychic nature in general." Objects are there for us in general and only there through consciousness, more exactly through figurations of cognizing, perceiving, remembering, phantasizing, [153] then also thinking in higher, specifically logical figurations, up to systematic scientific cognition. Through these, we have a consciousness, first, of natural things; as being or presumably being we experience and think them, or as possibilities we hypothetically posit them, and in logical acts we strive to determine their true being and being so.[43] [154]

---

[43] <Inserted page: marginal note in the following text. This probably belongs in the insertion in the "Introduction to Ethics" on material sciences and normative science. Edmund Husserl, *Einleitung in die Ethik. Vorlesungen Sommersemester 1920 und 1924*, ed. Henning Peucker, *Husserliana 37*, (Kluwer Academic Publishers, Dordrecht, 2004, 259–320, forthcoming in English translation by Andrew Barrette; or, in "Introduction to Philosophy" Edmund Husserl, *Einleitung in die Philosophie. Vorlesungen 1922/23*, ed. Berndt Goossens, *Husserliana* 35. (Kluwer Academic Publishers: Dordrecht, 2002).>

It follows that the universality of the formal doctrine of objects, thus of all pure mathematics, is also in this respect an empty one since it does not speak of predicates that have their origin in the sphere of heart and will, and not of objects that from the outset have their sense content included, although the functions of heart also have a universality and, in the opposite direction, encompass all objects of knowledge; for, after all, no object is thinkable that could not in some respect, at least indirectly, move the heart, even if it is only in the way in which the formal mathematical theories have their greater or lesser beauty. Marginal note to <p.> 126ff. <= pp.152ff.>.

Accordingly, it can also be argued with good reason that the same universality of validity that the formal theory of objects, mathematics, and formal logic have for all actual and ideal possible worlds insofar as they are worlds of objects at all, existing in the sense of knowledge, the parallel normative disciplines and formal sciences of value-objects and practical objects for the actual and all possible worlds insofar as they are not only worlds of cognition, but also worlds of values and goods, and they necessarily are so, above all every individual world of being. It is what it is only as environing world of a subjectivity. And a subjectivity, which we think of as rational subjectivity, is necessarily such a subjectivity, which not only finds itself in relation to a nature in the sense of a mere nature, but which creates a spiritual environing world for itself in the special sense. And

If we are already conscious of objects in some manner, as intuitively there or already theoretically known, they also touch and determine our heart. They are valuated as that which they are, be it in the manner of self-values, be it through taking on value characters mediately, through transfer. Then, however, the function of cognition is set in motion again; the value characters, which accrue to the objects in valuing, are drawn into the sense of objectivity in the form of properties, value predicates. The relevant object consciousness has been enriched, and we have objects of a higher level which are conceived from the outset as endowed with value characteristics, as, for instance, beautiful things. And likewise with regard to practical functions and corresponding practical predicates. Objects are there from the outset as works, as tools, as works of art, and so on. We spoke of this already when we discussed the idea of culture. Here we only emphasize that, in accordance with the essence of an I-subject, it is a matter of an endless process according to which ever-new formations of objects can be brought about, in which always new predicates of value and predicates originating on the part of the will can arise precisely because the valuative function and the function of the will connected with it can *a priori* come into play again and again, while, step by step, cognition brings new object-based predicates to apprehension. Objects always stand under the formal categories of objectivity, and the value predicates or practical predicates included in the content of the objects of cognition are material from the logical point of view. On the other hand, the valuing functions and practical functions are always ready for the accomplishment of new object-based sense-bestowal; and these functions have their most general typology, that is, their form, which corresponds to the pure essence of valuing in general and of the practical in general, and which apart from all particularity in which they, determined by already given object-based features, hold sway. [155]

---

precisely this leads us to the ultimate decisions concerning normative and factual sciences. *Crossed out*: Sciences like set theory, number theory and the disciplines of other formal mathematics arise in such a way. In them, likewise, establishment of norms is not the specifically scientific theme. The theme is the universe of possible objects as objects of possible determining truths in general. Also, from mere formal mathematics, one cannot derive any material science, no science of determinations domains of objects that are originally given only by intuition, although, conversely, mathematics can helpfully be applied in all such regions. Cognition, as positing objects and judicatively determining objects is universal, it also encompasses the sphere of heart and will and the values and value contents, substrates of will and determinations of will, and so on, constituted here, namely, what, for example, has constituted itself as value in the object in a value consciousness, this is comprehensible for cognition in the form of a value object. In the determining cognition, the character of value can be found as a predicate in the otherwise already given object and be attributed to it determinatively. Each step in the valuative and practical action creates at the same time for the judging subject comprehensible and determinable objects of a higher level; similarly, each step of judgment creates new objects: the judgment contents or states of affairs to be extracted from the judgments. The determining cognition must already, as we can see, have a substrate, objects that are pre-given to it. But these objects can also be pre-given to it by the fact that objects that are already pre-given elsewhere affect the heart in a certain way, and, as a result, receive a value coloration; and now the object to which the value character adheres can itself enter into determinations as a substrate by affecting the judging subject in its value concretion, with cognitive movements to follow.

Precisely that which belongs to the most general essence of valuing and willing, of the choosing decisions, expresses itself, cognitively grasped, in the universal *a priori* axioms of the doctrine of values and the doctrine of practical goods and oughts. Just as the logical axioms express what applies to objects in general, is valid for truth and being in general, for cognition and correct cognition in general, regardless of what may determine the particular sense content, the features of the content of the objects, so too do the formal axioms of value express what applies to values and correct values in general and to supposed and true values in general, regardless of what objects there are that are valued and what factual content in these objects becomes the substrate of the valuation. Likewise, from a practical point of view, the *a priori* has a formal and thus most general meaning, namely, one that concerns willing in general and goals of willing in general, ends and means in general, the absolute ought in general; this is so regardless of what the more specific types of objects, the persons (ego-subjects), the circumstances, the particular determining valuations are that can be taken into practical consideration.

The universality of the valuative and practical functions, despite their foundation by already preceding cognitive functions, is no less than the universality of those very cognitive functions, and entails that the most supreme formal sciences of principles, logic, the formal theory of values, and the formal science of practice mutually overlap in a wonderful manner and are thereby referred back to one another. (We call these: formal axiology, formal practice.)

1) Logic as a formal theory of science in the sense defined earlier encompasses all sciences in general insofar as it establishes universally valid laws for all sciences as such. Thus, it also encompasses in its validity the formal theory of value and formal practice and, at the same time, it encompasses itself. It corresponds to its own norms.

<2)> Secondly, formal science of values (formal axiology) embraces all value areas in their formal generality; since all sciences and, in them, all truths are values and in scientific striving are constantly regarded as values, the formal science of values, in this way, embraces all sciences, not, of course, as logic does, but precisely as values. In this way, however, the science of values is itself a value and an infinite system of values and embraces itself. [156]

3) Likewise, for ethics as formal ethics or formal practice: all sciences are not only values and systems of values, but they are also practical construction; and formal practice as the science of the *a priori* of every possible practice in general and of every possible practical construction and its practical values thus embraces all sciences and, thereby, also itself. It can also be said: cognitive, valuative and practical reason interpenetrate one another, and their essential lawfulnesses overlap one another without interfering with one another and without somehow forfeiting their particular peculiarity.

(A logical law encompasses a value law logically only insofar as the latter is a law, but not insofar as it speaks about values. For axiology as such speaks of values and laws for values, not of laws for truth and being in general; also, vice versa, a law of values does not axiologically encompass a logical law, as if a logical law were a statement about values of the logical sphere; since logic does not speak of values at all, the concept value does not occur in it at all. But truth is also valuated and can accordingly be brought under value predicates. As soon as this happens and statements are made about it, we stand in the value-scientific sphere.)

Like the supreme essential forms of reason, there are its three correlative concepts of truth: logical truth, axiological truth, and practical truth; and likewise: object-based being in general, value being in general, and the being of the ought in general; these are in the curious relation of coordination and yet of interpenetration, which, of course, is to be understood with some caution. But we must not linger here, for the following will elucidate the significance of such insights for a highest doctrine of categories.[44] [157]

In our earlier lectures, we formulated a large number of axioms for the world of values and the practical world, which were almost entirely of the character of formal axioms. I remind you only of the law of absorption and all the axioms culminating in the law of the categorical imperative, which is itself the highest law in formal ethics. In any case, the idea of a formal axiology and (what is nearer to us in our preferential interest in ethics) the idea of a formal ethics has been sharply clarified for us. Of course, in formal ethics, we must not speak of human beings, the species *homo* on this earth, but only the formal idea of an acting I as such, functioning in it as subject and this, not as related to factually given nature and the human world, but to an environing world as such in formal generality. However, then there is nothing standing in the way of materially restricting the formal *a priori* and, still in eidetic generality, of devising an *a priori* theory of norms and a technique of good will and action for humans, namely, in such a way that it is bound to the idea of the human and thinks in relation to humans of a human environing world, which, for its part, would itself have to be grasped in a pure idea. Although such an *a priori* human ethics naturally stands behind formal ethics in terms of essential generality, it is nevertheless a clear and thoroughly indispensable postulate.

Historical ethics, in its ever-new attempts from antiquity to the present, wanted from the outset to be a human ethics, it wanted to be a technique that would give human beings general rules concerning how they could best shape their practical life. Of course, some of the rules that were brought to expression were of *a priori* and even formal validity, though they were related to the human being in their earthly circumstances. Even after the Platonic discovery of the *a priori,* it took millennia until the idea of an *a priori* ethics had come to pure formation, not to mention until the idea of a formal ethics as a parallel of a formal logic was put forward in its

[44] *Marginal note*: Important supplement, unfortunately forgotten in the lecture. Supplement: Ethics thus encompasses the universe of all sciences and arts and individual practical achievements in their rule-giving and thus, in the backwards reference that is essentially peculiar to highest disciplines, itself. Since the ethical circumscription is called "practical demand," this indicates: By developing the demands of the categorical imperative systematically, ethics, namely, *a priori* ethics, also comes as to a practical demand to the conclusion that ethics should be developed. Among the scientifically deduced demands that it establishes, there is also the one that all of these demands and also it itself should be established scientifically. This is demanded by the categorical imperative. By way of this self-demand of ethics, the demands of all genuine sciences and arts come as ethical demands.

peculiarity and significance. It comes into consideration here that, as we have heard and will still hear, it required tremendous efforts and great intellectual struggles before the Platonic *a priori* had gained its pure sense, freed from all mystical admixtures, or had asserted itself in the purity of its validity.

How already in the primitive beginnings, the *a priori* was apparent and yet unnoticed in its peculiarity, I will show by way of hedonistic ethics and, at the same time, link to it a supplementary elaboration on the idea of ethics itself, [158] which is indispensable for our philosophical interests.[45] Ancient ethics begins as a doctrine of happiness, and the ethical question in general remains bound to this form in antiquity: How does one gain one's *eudaimonia* and what determines it? All human beings, indeed, all living beings, strive for happiness, the pursuit of happiness belongs to the nature of humans. What is it that makes human beings happy, how is their true happiness determined?

A very simple answer to this question is given by hedonism, which appears in its first and in principle completely pure form in Aristippus. He called himself a student of Socrates and was indeed one of his regular companions, but in spirit, he was lightyears away from him. Aristippus's clear answer to the question of *Eudaimonia* is as follows: striving for happiness is striving for pleasure. All living beings strive for pleasure and flee pain. Only in pleasure is the striving fulfilled. But not every pleasure is equal to every other one, human beings prefer the higher pleasure by nature. The human being therefore finds their true happiness, or at least the greatest possible happiness that can be attained through their own activity, when they partake in the greatest possible pleasure and the least possible displeasure. Of course, it must be considered that the attainment of pleasure may have partly pleasurable, partly painful consequences in the future. All this must be taken into account and practical reason, φρόνησις, demands of the human being that he be the constant calculator of good pleasure, such that he solves for every day and for their whole life the maximum problem correctly, namely, that of the greatest possible pleasure.

But what does higher and highest pleasure mean? The answer according to Aristippus is the following: the quantitatively greater, the more intense, the longer lasting pleasure. Furthermore: several pleasures, which are all at once realizable, combine in accordance with their quantities into a pleasure quantity; and, in the sum, on the other hand, pains, which are inevitably connected with the pleasures, function as negative quantities. If we now separate sensible and spiritual pleasure, this is, according to Aristippus, only a difference of the objects: sometimes we take pleasure in wine, at another time, in a tragedy or a marble statue. Those who call the latter pleasures "higher" consider their quantity of pleasure to be greater. Of course, according to Aristippus, he errs there, because, as he seriously claims, sensible pleasure is the highest, it has the greatest quantity and is also attainable in the greatest quantity. So, let us become pleasure-seekers! This is what "practical reason" demands. [159]

[45] *Marginal note*: Formal character of hedonistic ethics.

But where is an *a priori* here, you will ask? And should it even be insinuated that this common doctrine of pleasure is *a priori* legitimated? No, not by any means. But for the time being, we could already point to the guiding idea that shines through it that the human being, and then obviously every subject of will in general, is of course practically reasonable only if he realizes the best possible. But apart from that, suppose the hedonist was right with his fundamental principle, "pleasure is the good, the good is pleasure," right with the principle, "the human being cannot strive for anything else than pleasure," and, of course, by this is meant their own pleasure. And suppose that all pleasure is quantitatively comparable with all pleasure. We would then have exclusively homogeneous quantities in our practical realm.

In order not to appear to regard the doctrine of Aristippus as worthy of consideration, we can also choose the hypothesis that, sometimes in our practical realm, only pleasure values and quantitatively comparable pleasure values come into question. As, for instance, the case where no higher good is at stake than that of how to satiate ourselves, and now the menu is handed to us for the choice of the best. Now, obviously the quantitative principle is valid, and we will no longer have reason to doubt that it is *a priori*. For this is evident: where homogeneous pleasure values and values in general allow a quantitative comparison in some respect (and, obviously, only homogeneous ones allow it), the greater quantity is the reasonable one to choose: the roasted goose as opposed to sausage or sausage substitute and the roast goose twice versus once, if the wallet does not protest, of course. This is an evidence of practical reason.

Of course, it is also valid in other spheres that are higher in our opinion. If our capacity for attention is great enough, the longer concert is preferred over the shorter one, if only for the sake of quantity. More words, more musical performances combined already give a higher value for quantity reasons alone. Of course, quantity alone does not make it so, since the artistic effects do not merely accumulate, but can subsequently combine to form an increased, and in other cases, diminished unity of value. But if this is not the case, quantity alone decides. This *a priori*, then, is certainly contained in hedonism; and on such an occasion we recognize that, in *a priori* ethics, its own principles of value-quantity, principles which obviously have a purely formal character must also be established.

It need not be said that these principles cannot play an exclusive and somehow dominating role, since it is evident that mere summation is an abstract limit case. [160] Also, as quickly became apparent to us, values that appear together in the unity of a consciousness merge into wholes, such that, with the extension of the duration and the temporal addition of new goods and likewise with the increase of their intensity, changes in the combined total value that cannot be calculated according to principles of summation will go hand in hand. For the rest of it, we need not dwell on the criticism of Aristippus's doctrine, since it is, from the outset, beyond doubt that already the principle of the quantitative comparability of all pleasures and pains is fundamentally wrong: it proceeds as if we could convert the value of a symphony or a heroic deed, even if we wanted to misunderstand it as pleasure value, into so and so many pounds of roasted goose or caviar. Hence, even the hedonism of the later Cyrenaics, which is higher with respect to its sentiment, and which places the spiritual goods above the sensible, is not any better in principle, since it only judges that the spiritual goods are quantitatively more favorable, that they last longer, and so on.

Of more importance for us is the question as to whether the ethics of happiness is not correct, after all, if its principles are appropriately supplemented; whether we have not made a mistake in our general determination of the idea of ethics by not starting from the idea of happiness, which is so readily available. Is not all striving and choosing directed toward satisfaction, and does not "the best among the achievable" have, from the outset, the sense of "the greatest possible satisfaction," only that the expression "greatest possible" must not be understood in a literal, merely quantitative sense? Without entering into longer and more detailed discussions, we can nevertheless quickly gain clarity in the main issue.

Without doubt, it belongs a priori to the essence of every willing that its fulfillment as realization of the goal of the will is *eo ipso* pleasurable fulfillment. The satisfaction of the will, its discharge as fulfillment, attainment is *eo ipso* also an experience of pleasure. But as Aristotle already saw in his famous criticism of the hedonist Eudoxus, one must carefully distinguish between the two. And hence one must not confuse the goal of the will and the satisfaction of pleasure that necessarily occurs in successful action. Certainly, we can also choose pleasure as the goal of the will, and we do it often enough. But even then we have to distinguish: the pleasure that is our goal and the pleasure of satisfaction of attaining this pleasure, even if it fuses into a unity with the attained pleasure. On the other hand, we do not always strive for pleasure, and it is even wrong to think that every striving must necessarily have pleasure as its goal. [161] The seductive aspect of this popular opinion, which has captivated all hedonists up through Mill, Bentham, Fechner, has its root in the fact that, looking from the outside at the phenomenon of the attainment of the will, one always found in the attainment, even as a necessary component, the pleasure of the attainment, and that one did not see, empathizing from within the striving and acting, that the goal of the will itself is only exceptionally a pleasure but that at all times it is that which is supposed to be a value and is posited as a value to be realized. If we value science practically, it becomes our goal; this is likewise with artistic creation and so many other things. To posit pleasure, a determinate pleasure or an indefinite general pleasure, as a goal, is to attach value to it and to want to realize it. It can be sensual pleasure, it can be some higher pleasure, for instance, satisfaction in the delight of a work of art, or it can be indeterminate, for instance, in the sensation of some sort of delight, just this indeterminacy. But then there is a clear difference.[46] Striving for scientific knowledge is not the same as striving for pleasure in scientific knowledge. In the one, the knowledge itself is the final goal, in the other it becomes the means for the pleasure of satisfaction expected from it. And then, as already touched upon, there must arise as success of the realization of this satisfaction of pleasure a second satisfaction of pleasure.

[46] *Marginal note*: Pleasure as sensual excitement and mood, caused by fulfillment of the striving—and fulfillment of the striving, in which the wanted and ought is attained as "true" goal, that which is an ought for me and the ought-to-be itself as truly good. That is another difference.

Because of the functional connection between the goal of the will and the pleasure of satisfaction, it is then clear that, in a certain sense, the ideal of the greatest possible *eudaimonia* and that of an ethically best life go hand in hand with each other.[47] First of all, it can be said: whoever does the ethically good, their will is thereby the most fulfilled as is thinkable, they attain their best possible goal of will.[48] But even if one were aiming at pleasure, including all possible pleasures of satisfaction, one could not conceive of a higher pleasure as a goal than the one which,[49] in the attainment of ethical satisfaction, would shine before one as the blossom of consistently ethical actions. It is, in an evident manner, elevated in rank above all achievable pleasure values. But of course, as soon as we ask why—Why is the ethical life the most satisfying, the happiest one thinkable?[50] [162] Then we quickly recognize that the happiness value decisive here is not equal to the ethical value, but is derived from it, and we recognize that no other value can become decisive at all. For no other reason, however, is the ethical satisfaction the pleasure of highest dignity than just because it derives its dignity from the highest value, the ethical. Practically, I must not prefer any other joy before this ethically awakened joy, because otherwise I would prefer a bad one, I would violate the categorical imperative. Admittedly, I also see that I can realize this joy only if I do not think about joy, but about the goods of all kinds that can practically come into consideration for me and pursue their inherent value. I see then that *eo ipso* this most beautiful joy, highest in dignity, must blossom for me. Furthermore: extra-practically, no other joy can have a higher value, because if it had it, I would be allowed to strive for it in contrast to the ethical joy. And that would result in an ethical contradiction.

We have (thus chosen the correct way and) recognized that the principles of ethics must not be determined by the idea of happiness and the most satisfying life possible, while, on the other hand, ethical life still proves to be the most satisfying thinkable that is possible for the willing subject and its praxis. At the same time, that popular sophistry that states: "everyone can do nothing else than what gives him pleasure; even those who practice neighborly love do it only because it gives them pleasure to help others" can no longer have any power for us. Such sophistry states as if there were no pure love for the other, which thus has its *telos* in the other, and as if the joy in the other and the act of love for the other, because it is an experience of joy and activity in us, were just joy in our joy and striving for our own benefit.[51]

It is similar with the ethics of perfection, which often appears interwoven with the ethics of happiness. It looks to a completely different aspect in the multifaceted essence of ethical life. While the ethics of happiness looks toward the sense of

---

[47] *Marginal note*: It does not work that way.

[48] *Marginal note*: Not always objective! Even if subjective: That which one should choose, that which one chooses and does.

[49] Pleasure. Crossed *out, and a marginal note included*: Satisfaction!

[50] *Marginal note*: But that still does not make happiness.

[51] *Marginal note*: But does joy, pleasure always play a role?

satisfaction attached to the *telos*, the ethics of perfection looks toward the I from which the will proceeds and which, depending on the direction of the will and the value of the will, elevates and degrades itself. [163] There is nothing standing in the way of our outlining the idea of a normative and practical discipline of the self-perfection of a personality; and, as soon as one considers it, one immediately recognizes that it would have to be carried out mainly with the ethics of the categorical imperative, the ethics of the good will, while it can only be legitimated as a derivation from the ethical doctrine of principles that we have obtained. It is clear that, for every subject, the doing of the ethically required best would be at the same time the best for their self-perfection, as well as that the idea of the best possible perfection gained by self-activity would have to include all ethically required goals in itself.

It must be taken into consideration here that the question concerning how persons, whether correctly or incorrectly, value and that for which they practically preferentially decide is by no means irrelevant for the valuation of the personality itself.[52] On the contrary, the value and disvalue of personalities is derived from this in many ways. It is, after all, *a priori* evident that a person who, with insight or in the practical certainty of being able to restore the insight at any time, values a truly beautiful thing as beautiful, a truly good thing as good, is, by this very fact and in this respect, truly good and all the better, the more perfect the good thing is. It can also be said: not only is insight and the rational conviction derived certainly from insight better than blind opinion, but the one with the insight and rational conviction is better (as such) than the one blindly opining.

Furthermore: an act of will motivated by insight, which "follows the insight," is better than one that does not follow it. Also, however, the corresponding willing (that of the insight following subject itself) is better than the subject following a blind opining. It is the same in preferring too: practically preferring that which is recognized as better also gives value to the subject, whereas one becomes unvaluable as soon as one prefers a thing recognized as inferior to a higher one in the choice. Yes, here even the conviction as such is sufficient.

It is evident and always *a priori* that a person, insofar as they make a general resolution of will to shape their whole future life in the sense of the categorical imperative, is better than if they merely resolve to do the best possible in the particular case without having thought of and without thinking of a general self-legislation. [164] With respect to such *a priori* cognition, it can be asked: When does a person gain for himself the highest practical good value? By doing good a person perfects themselves *eo ipso*. A person can recognize that and, above and beyond this, want to perfect themselves. Each act establishes for them *eo ipso* a habitus, every valuable act a relatively valuable habitus that ennobles them. Now, however, habit and practice also practically come into question. Although habits indeed seem or characterize blind inclinations, they can nevertheless also be allies in the striving for the good, especially insofar as they can also, as habitual inclination, following the insight into the value of choosing the best, facilitate the doing-good and thus

[52] *Marginal note*: Axioms.

counteract other blind inclinations that so often lead to the decision for the worse, even though it is recognized as worse. (Indirectly, then, the formation of habitual inclinations and, as the case may be, combating them, insofar as they are within our practical power, also serve self-perfection and are a theme of ethically required legislation.)

We do not need to go further. It is clear that all such considerations in the sense of ethically directed self-education, the aim of which is the greatest possible self-perfection, has its right. But, on the other hand, it is also clear that all the practical values that result here are derived from the idea of good will and the categorical imperative, and that a life carried out according to this principle and according to the associated axioms brings with it, *eo ipso*, self-perfection; although, from such principles, the limits must be determined within which it is useful and then also necessary to posit oneself the goal of self-perfection explicitly, and it also must be determined how it is to be investigated which particular spheres of the soul, like the training of favorable inclinations, require a special cultivation. It can also be said: Self-perfection is a necessary theme of every empirical as well as *a priori* ethics. But this idea does not designate the central concept of ethics, it leads us back to more primitive concepts, to the formal axiological and practical categories and the corresponding axiological principles. However, this whole consideration has also enriched us to the extent that we have become acquainted with quite a number of new axioms, which can give us an idea of the richness of the ethical *a priori*.

Only one important supplement should now be added, which at the same time should serve us as a transition to social ethics. [165] Formal ethics construes an ideal for self-perfection: the formal ideal of the ethically most perfect personality. This ideal is now essentially related to the ideal of a most perfect personality *simpliciter*, but by no means do both ideals coincide. Everyone says to themselves as an ethically reflecting subject: I should do my best, that which is for me the absolute ought. And likewise: I should become as good as I possibly can, I should become an ethically perfect personality. But, first of all, it should be noted that this talk of an ethically perfect personality has its ambiguities. I have ethical perfection in one sense when I do in unfailing consistency what according to my conviction is the best among what is achievable. This is a practical idea, the idea of an infinite task in view of the constant struggle with tendencies of decline; yet, it is a possibly achievable ideal. But a further ideal is that I value and choose in constant consistency, always on the basis of insight and realize the best in accordance with perfect insight. I can never hope to reach this ideal, since my insight will undoubtedly let me down often enough. But an ideal task is designated, provided that I know that, within certain limits and progressively, insightfulness in judging and valuing is in my power. I must therefore take this up in my will, and the imperative of self-perfection formulated above then demands, *eo ipso*, the closest possible approximation to this second ideal, which lies in the infinite.

If we now compare several personalities, each of which we think as corresponding to the first ideal, they fulfill the formal imperative demand to be as good as they can be in general, to shape themselves in as a good a way as they in each respective case in general could. For this reason, however, the personalities need not be on the same level of approximation to the second ideal, as when, for instance, one is intellectually limited, the other visionary, more gifted, more capable of development. Everyone stands under the categorical demand of the ethical imperative, and each has their ethical perfection when they consistently fulfill this demand. But everyone is only who they are and not someone else, and what the one can do, the other cannot yet do; values that are open to one are not open to the other; and even if they are within one's realm, one may not incapable of properly weighing them, while the other may well be able to.

In ethical perfection in the proper sense, all are equal who act in consequence of ethical sentiment, who shape their personalities into the best they can become according to their dispositions and under the circumstances of their environing world. [166] Each of these human beings has the same ethical perfection, because ethical perfection by its very essence is only the expression of the fulfillment of a formal demand and is related to the particular person.[53] *A priori*, it always takes on new content depending on the person. On account of fulfilling of this ideal, humans have dignity or, better, the highest possible dignity, for they have dignity already on account of an ethical sentiment predominating in them, and a dignity all the more perfect, the closer this predominance comes to absolute predominance in the sense of the ideal. "Dignity," then, is properly the value predicate belonging specifically to the personal subject as the subject of categorically demanded action, and as a predicate adhering in the pure form it is not affected by differences in the content of the person.

This means, however, and in accordance with that which was seen above, that with the ethical valuation of a person as a formal one, a different good valuation, which is related to the factual stock of the personal qualities, of some other person cannot be confused. As soon as we take into account the peculiarities of the personalities, we find infinitely many value differences also between equally ethically perfect persons. A Kant and a farmer, both thought of as ethically perfect, both constantly doing the best according to their best knowledge and conscience, are ethically equal to each other, the same ethical admiration is due to them. And yet there is an enormous difference in value between the two. Now, we do not value the persons ethically, but fully in terms of content, like other objects, as good values: the thinking, feeling, creating of the philosophical genius and the personal qualities, which live themselves out in such acts, stand on an overwhelmingly higher level compared to those of the simple farmer. Of course, however, it would be fundamentally wrong and even ethically reprehensible if the bearer of the higher value level, after this extra-ethical value comparison, would take credit for their height and look down contemptuously on the lower one.

However, regarding the value comparison of persons here called "extra-ethical," which is directed to the good values of the persons and has to take into account all their valuable aspects, it is to be noted that ethical perfection also plays its role, and a very curious one. Let us leave aside all values of means, of which there is no

[53] *Marginal note*: Dignity.

question here. [167] It is a question of the intrinsic value of a person, apart from the fact that he can become a value as a means like anything else. First of all, we have to say: in the ethical valuation, we consider only the ethical sentiment. In the good valuation of the person as such, where we must consider the whole person, then the ethical sentiment is a mere component. The ethical character is not the whole character, but only one aspect of the personality; as valueless it also determines the good value of the persons, as good as the other valueless sides of the person. But the matter is not so simple, for that is precisely the curious thing, namely, that all other values, for instance, those that characterize the genuine artist, the scholar, the educator, are not ordered in the same way as the ethical value, but rather that the ethical character first of all ultimately determines the value of all other aspects of the character; or also that, on the one hand, it is a component of the overall character before any question of value and disvalue, but, on the other hand, it is to be regarded in the valuation as a form that co-determines the value of all other components of character; and this is so in such a way that every character trait is to be attributed to the intrinsic value of the persons only to the extent that it is imbued with ethical sentiment and in the degree to which it is so.

Certainly, the enormous scientific ability and achievements of a Plato, Kant, of a Gauss, or of the artistic ones of Michelangelo or Goethe are tremendous values. But we must separate the value of the scientific formations, of the works of art, and so on, which is value in itself, whatever the value of the personalities who have produced them, and, on the other hand, the value of these very personalities themselves. As producing values, they are themselves valued, but this value is a mere transference value, if the personalities have not created in an ethical sentiment, if they have created, for example, not in order to achieve great and beautiful things, but in order to become famous and honored. So, if we want to determine the value of a person, namely, the pure intrinsic value of the person and not the person's value as a mean and as a useful value for others, then their ethical character is the fundamental value from which all other aspects of character that can be exhibited in the person as positive values assume their value, which is to be attributed to the value of the person themselves.

In other words: we may honor and admire the genius only if they philosophized, mathematized in an ethical spirit; regardless of whether, in abstract consideration, the brilliant achievement is a great value, it cannot be attributed to the person themselves without further ado as their intrinsic value. The vain, the unethical have their reward there. Only the ethical ones, precisely because they strive for nothing for the sake of a reward, receive the highest gift of loving admiration in return. [168] (All values to be called "personal" are thus, indeed, real components of the value of the person only when they have assumed the form of the ethical sentiment or, included in it, have experienced their ethical determination from it.) Basically, all personal valuation is indeed ethical valuation, only that we must separate the purely ethical valuation, which only relates to the form of the sentiment, and the concrete good valuation of the person, which, by determining their personal intrinsic value, finds in the ethical form the condition of the possibility of the intrinsic value of the person.

(In the idea of a most perfect personality, which lies in the infinite, the idea of a most perfect ethical personality is included. In the sense of the latter is the fact that all their character traits have taken on a purely ethical form through the constantly exercised consistency of the ethically good will or through the discipline of ethical self-education, and thus never appears and never can appear in any other way than in the service of the ethical sentiment. Every ethically directed personality is directed toward this purely ethical forming of its character. The ideal of the most perfect personality as such lies beyond this, insofar as materially different purely ethically formed characters can be of very different value according to the content which is ethically formed there.)

The question of self-perfection and the related questions of valuation of the person lead us to the question of egoism and altruism, and so, too, to the question of the relationship between ethics and morality and between individual ethics and social ethics. If questions like "What should I do?" and "How do I become good?" are basic ethical questions, they are addressed to a respective particular person. And what demands arise in responding to them are no less demands on the particular person.

Now, one could object: in that it demands self-perfection from me, ethics calls me to egoism; I should further my own good. And is this not to be looked down upon as egoism? However, let us play no games and consider the concept of egoism! Egoism is every (habitual) preferring of one's own over the alien, more precisely, every preferring of a really or supposedly good thing in my sphere of action while disregarding, damaging or not considering the practical goods in the sphere of action of another. Everyone has a certain practical realm, which is directly their own and that would remain for them even if all others are disregarded. By virtue of empathy, however, each I also insight into another I and its practical sphere of goods, [169] and through this already, and even more so by virtue of the reciprocal communicative understanding through which social community emerges, the practical sphere of action of one extends beyond itself into the spheres of action of the others. The ethical *a priori*, however, demands that the same practical goods with the same values are to be applied in the ethical measure. Wherever a good for the other belongs also to my sphere of effects, it is, *ceteris paribus*, equal to the good for me; then, under the guidance of the categorical imperative, it is to be decided from the circumstances when I should set aside my good or give it preference, just as I otherwise look for and find reasons for preference in the choice of two goods that are, according to their proper value, equal and incompatible with each other. I am egoistic only when I do not care about this rule and, without further ado, disregard the good for the other, where I do not even consciously set it aside.

With egoism goes hand in hand the intellectual self-love and self-evaluation. This has its counterpart in the genuine human love as genuine love of neighbor and self-love, meant as a valuing and practical comportment that is purely motivated, determined by the intrinsic personal value and, in it, therefore and above all, by the ethical value. In willing to be good and willing to become good, I show the right self-love towards myself; just as I show the right love of neighbor by attending to and fully recognizing the good in the other and by further doing my best, within the framework of my life's task, to support them in the fulfillment of their ethical tasks, and thus also in their self-perfection. As is well known, this is where the greatest practical difficulties lie. The norms and practical rules concerning the right comportment towards others, thus also those relating to the rejection and practical overcoming of egoism, are the so-called moral norms in the specific sense. Obviously, the moral in no way exhausts the ethical in general. Ethical reason is practical reason as such in its full breadth and principal unity; moral reason is reason in practical behavior toward others.